In
Other Worlds

By Margaret Atwood

FICTION

The Edible Woman

Surfacing

Lady Oracle

Dancing Girls

Life Before Man

Bodily Harm

Murder in the Dark

Bluebeard's Egg

The Handmaid's Tale

Cat's Eye

Wilderness Tips

Good Bones

The Robber Bride

Alias Grace

The Blind Assassin

Good Bones and Simple Murders

Oryx and Crake

The Penelopiad

The Tent

Moral Disorder

The Year of the Flood

POETRY

Double Persephone

The Circle Game

The Animals in That Country

The Journals of Susanna Moodie

Procedures for Underground
Power Politics
You Are Happy
Selected Poems
Two-Headed Poems
True Stories
Interlunar
Selected Poems II: Poems Selected and New 1976–1986
Morning in the Burned House
The Door

NONFICTION

Survival: A Thematic Guide to Canadian Literature
Days of the Rebels 1815–1840
Second Words
Strange Things: The Malevolent North in Canadian Literature
Two Solicitudes: Conversations *(with Victor-Lévy Beaulieu)*
Negotiating with the Dead: A Writer on Writing
Moving Targets: Writing with Intent 1982–2004
Writing with Intent: Essays, Reviews, Personal Prose 1983–2005
Payback: Debt and the Shadow Side of Wealth

FOR CHILDREN

Up in the Tree
Anna's Pet *(with Joyce Barkhouse)*
For the Birds
Princess Prunella and the Purple Peanut
Rude Ramsay and the Roaring Radishes
Bashful Bob and Doleful Dorinda

Nan A. Talese

DOUBLEDAY

New York London Toronto Sydney Auckland

In Other Worlds

SF and the Human Imagination

Margaret Atwood

 Published in the United States by Nan A. Talese / Doubleday, a division of Random House, Inc., New York.

www.nanatalese.com

DOUBLEDAY is a registered trademark of Random House, Inc. Nan A. Talese and the colophon are trademarks of Random House, Inc.

Simultaneously published in Canada by Signal, an imprint of McClelland & Stewart Ltd., Toronto.

Book design by Maria Carella
Jacket design by Michael J. Windsor
Jacket photograph by Alex and Felix. www.alexandfelix.com

Library of Congress Cataloging-in-Publication Data
Atwood, Margaret, 1939–
In other worlds : SF and the human imagination / Margaret Atwood. — 1st U.S. ed.
p. cm.
Includes bibliographical references.
1. Atwood, Margaret, 1939—Knowledge—Literature. 2. Atwood, Margaret, 1939—Knowledge—Science fiction. 3. Science fiction—History and criticism. 4. Science fiction—Authorship. I. Title.
PR9199.3.A8Z545 2011
813'.54—dc22 2011013776

ISBN 978-0-385-53396-6

PRINTED IN THE UNITED STATES OF AMERICA

10 9 8 7 6 5 4 3 2 1

First United States Edition

For Ursula K. Le Guin

Contents

Five Tributes

Appendices

In Other Worlds

Introduction

I'm a fifty-three-year-old writer who can remember being a ten-year-old writer and who expects someday to be an eighty-year-old writer.
OCTAVIA BUTLER

I*n Other Worlds* is not a catalogue of science fiction, a grand theory about it, or a literary history of it. It is not a treatise, it is not definitive, it is not exhaustive, it is not canonical. It is not the work of a practising academic or an official guardian of a body of knowledge. Rather it is an exploration of my own lifelong relationship with a literary form, or forms, or subforms, both as reader and as writer.

I say "lifelong," for among the first things I wrote as a child might well merit the initials SF. Like a great many children before and since, I was an inventor of other worlds. Mine were rudimentary, as such worlds are when you're six or seven or eight, but they were emphatically not of this here-and-now Earth, which seems to be one of the salient features of SF. I wasn't much interested in *Dick and Jane:* the creepily ultra-normal characters did not convince me. Saturn was more my speed, and other realms even more outlandish. Several-headed man-eating marine life seemed more likely to me, somehow, than Spot and Puff.

Our earliest loves, like revenants, have a way of coming back in other forms; or, to paraphrase Wordsworth, the child is mother to the woman. To date—as what I am pleased to think of as an adult—I have written three full-length fictions that nobody would ever class as sociological realism: *The Handmaid's Tale*, *Oryx and Crake*, and *The Year of the Flood*. Are these books "science fiction"? I am often asked. Though sometimes I am not asked, but told: I am a silly nit or a snob or a genre traitor for dodging the term because these books are as much "science fiction" as *Nineteen Eighty-Four* is, whatever I might say. But is *Nineteen Eighty-Four* as much "science fiction" as *The Martian Chronicles*? I might reply. I would answer not, and therein lies the distinction.

Much depends on your nomenclatural allegiances, or else on your system of literary taxonomy. Back in 2008, I was talking to a much younger person about "science fiction." I'd been asked by the magazine *New Scientist* to answer the question "Is science fiction going out of date?" But then I realized that I couldn't make a stab at the answer because I didn't really grasp what the term *science fiction* meant anymore. Is this term a corral with real fences that separate what is clearly "science fiction" from what is not, or is it merely a shelving aid, there to help workers in bookstores place the book in a semi-accurate or at least lucrative way? If you put skin-tight black or silver clothing on a book cover along with some jetlike flames and/or colourful planets, does that make the work "science fiction"? What about dragons and manticores, or backgrounds that contain volcanoes or atomic clouds, or plants with tentacles, or landscapes reminiscent of Hieronymus Bosch? Does there have to be any actual science in such a book, or is the skin-tight clothing enough? These seemed to me to be open questions.

This much younger person—let's call him Randy, which was in fact his name—did not have a hard and fast definition of "science fiction," but he knew it when he saw it, kind of. As I told *New Scientist,* "For Randy—and I think he's representative—sci-fi does include other planets, which may or may not have dragons on them. It includes the wildly paranormal—not your aunt table-tilting or things going creak, but shape-shifters and people with red eyeballs and no pupils, and Things taking over your body." Here I myself would include such items as Body Snatchers—if of extraterrestrial rather than folkloric provenance—and Pod People, and heads growing out of your armpits, though I'd exclude common and garden-variety devils, and demonic possession, and also vampires and werewolves, which have literary ancestries and categories all their own.

As I reported in my *New Scientist* article, for Randy sci-fi includes, as a matter of course, spaceships, and Mad Scientists, and Experiments Gone Awfully Wrong. Plain ordinary horror doesn't count—chainsaw murderers and such. Randy and I agreed that you might meet one of those walking along the street. It's what you definitely would not meet walking along the street that makes the grade. Randy judged such books in part by the space-scapes and leathery or silvery outfits on their covers, which means that my speculations about jacket images are not entirely irrelevant. As one friend's child put it: "Looks like milk, tastes like milk—it IS milk!" Thus: looks like science fiction, has the tastes of science fiction—it IS science fiction!

Or more or less. Or kind of. For covers can be misleading. The earliest mass-market paperbacks of my first two novels, *The Edible Woman* and *Surfacing,* had pink covers with gold scrollwork designs on them and oval frames with a man's head and a woman's head silhouetted inside, just like valentines. How many readers picked these books up, hoping to find a Harlequin Romance or reasonable facsimile, only to throw them down in tears because there are no weddings at the ends?

Then there was the case of the former Soviet Union. No sooner did the Wall come down in 1989 than pornography flooded across the one-time divide. Porn had hitherto been excluded in favour of endless editions of the classics and other supposed-to-be-good-for-you works, but forbidden fruit excites desire, and everyone had already read Tolstoy, a lot. Suddenly the publishers of serious literature were hard-pressed. Thus it was that *The Robber Bride* appeared in a number of Soviet-bloc countries with covers that might be described as—at best—deceptive and—at worst—as a Eurotrash slutfest *in flagrante*. How many men in raincoats purchased the *Robber Bride* edition sporting a black-satin-sheathed Zenia with colossal tits, hoping for a warm one-handed time in a back corner, only to heave it into the bin with a strangled *Foiled Again!* curse? For the Zenia in my book performs what we can only assume is her sexual witchery offstage.

Having thus misled readers twice—inadvertently—by dint of book covers and the genre categories implied by them, I would rather not do it again. I would like to have space creatures inside the books on offer at my word-wares booth, and I would if I could: they were, after all, my first childhood love. But, being unable to produce them, I don't want to lead the reader on, thus generating a frantic search within the pages—*Where are the Lizard Men of Xenor?*—that can only end in disappointment.

.....

My desire to explore my relationship with the SF world, or worlds, has a proximate cause. In 2009, I published *The Year of the Flood,* the second work of fiction in a series exploring another kind of "other world"—our own planet in a future. (I carefully say *a* future rather than *the* future because the future is an unknown: from the moment *now,* an infinite number of roads lead away to "the future," each heading in a different direction.)

The Year of the Flood was reviewed, along with its sibling, *Oryx and Crake,* by one of the reigning monarchs of the SF and Fantasy forms, Ursula K. Le Guin. Her 2009 *Guardian* article began with a paragraph that has caused a certain amount of uproar in the skin-tight clothing and other-planetary communities—so much so that scarcely a question period goes by at my public readings without someone asking, usually in injured tones, why I have forsworn the term *science fiction,* as if I've sold my children to the salt mines.

Here are Le Guin's uproar-causing sentences:

> To my mind, *The Handmaid's Tale, Oryx and Crake* and now *The Year of the Flood* all exemplify one of the things science fiction does, which is to extrapolate imaginatively from current trends and events to a near-future that's half prediction, half satire. But Margaret Atwood doesn't want any of her books to be called science fiction. In her recent, brilliant essay collection, *Moving Targets,* she says that everything that happens in her novels is possible and may even have already happened, so they can't be science fiction, which is "fiction in which things happen that are not possible today." This arbitrarily restrictive definition seems designed to protect her novels from being relegated

> to a genre still shunned by hidebound readers, reviewers and prize-awarders. She doesn't want the literary bigots to shove her into the literary ghetto.

The motive imputed to me is not in fact my actual motive for requesting separate names. (If winning prizes were topmost on my list, and if writing such books would guarantee non-wins, my obvious move would be just to avoid writing them.) What I mean by "science fiction" is those books that descend from H. G. Wells's *The War of the Worlds*, which treats of an invasion by tentacled, blood-sucking Martians shot to Earth in metal canisters—things that could not possibly happen—whereas, for me, "speculative fiction" means plots that descend from Jules Verne's books about submarines and balloon travel and such—things that really could happen but just hadn't completely happened when the authors wrote the books. I would place my own books in this second category: no Martians. Not because I don't like Martians, I hasten to add: they just don't fall within my skill set. Any seriously intended Martian by me would be a very clumsy Martian indeed.

In a public discussion with Ursula Le Guin in the fall of 2010, however, I found that what she means by "science fiction" is speculative fiction about things that really could happen, whereas things that really could not happen she classifies under "fantasy." Thus, for her—as for me—dragons would belong in fantasy, as would, I suppose, the film *Star Wars* and most of the TV series *Star Trek*. Mary Shelley's *Frankenstein* might squeeze into Le Guin's "science fiction" because its author had grounds for believing that electricity actually might be able to reanimate dead flesh. And *The War of the Worlds*? Since people thought at the time that intelligent beings might live on Mars, and since space travel was believed to be possible in the imaginable future, this book might have to be filed under Le Guin's "science fiction." Or parts of it might. In short, what Le Guin means by "science fiction" is what I mean by "speculative fiction," and what she means by "fantasy" would include some of what I mean by "science fiction." So that clears it all up, more or less. When it comes to genres, the borders are increasingly undefended, and things slip back and forth across them with insouciance.

Bendiness of terminology, literary gene-swapping, and inter-genre visiting has been going on in the SF world—loosely defined—for some time. For instance, in a 1989 essay called "Slipstream," veteran SF author Bruce Sterling deplored the then-current state of science fiction and ticked off its writers and publishers for having turned it into a mere "category"—a "self-perpetuating commercial power-structure, which happens to be in possession of a traditional national territory: a portion of bookstore rack space." A "category," says Sterling, is distinct from a "genre," which is "a spectrum of work united by an inner identity, a coherent aesthetic, a set of conceptual guidelines, an ideology if you will."

Sterling defines his term *slipstream*—so named, I suppose, because it is seen as making use of the air currents created by science fiction proper—in this way:

> . . . I want to describe what seems to me to be a new, emergent "genre," which has not yet become a "category." This genre is not "category" SF; it is not even "genre" SF. Instead, it is a contemporary kind of writing which has set its face against consensus reality. It is fantastic, surreal

> sometimes, speculative on occasion, but not rigorously so. It does not aim to provoke a "sense of wonder" or to systematically extrapolate in the manner of classic science fiction. Instead, this is a kind of writing which simply makes you feel very strange; the way that living in the late twentieth century makes you feel, if you are a person of a certain sensibility.

His proposed list of slipstream fictions covers an astonishing amount of ground, with works by a wide assortment of people, many of them considered to be "serious" authors—from Kathy Acker and Martin Amis to Salman Rushdie, José Saramago, and Kurt Vonnegut. What they have in common is that the kinds of events they recount are unlikely to *have actually taken place.* In an earlier era, these "slipstream" books might all have been filed under the heading of "traveller's yarn"—stories like, for example, Herodotus's accounts of monopods and giant ants or medieval legends about unicorns, dragons, and mermaids. Later they might have turned up in other collections of the marvellous and uncanny, such as *Des Knaben Wunderhorn,* or—even later—the kind of You-won't-believe-this-hair-raiser to be found in assortments by M. R. James or H. P. Lovecraft or—occasionally—R. L. Stevenson.

But surely all draw from the same deep well: those imagined other worlds located somewhere apart from our everyday one: in another time, in another dimension, through a doorway into the spirit world, or on the other side of the threshold that divides the known from the unknown. Science Fiction, Speculative Fiction, Sword and Sorcery Fantasy, and Slipstream Fiction: all of them might be placed under the same large "wonder tale" umbrella.

. . . . .

This book is arranged in three parts. The first part, "In Other Worlds," is a personal history of sorts. Its three chapters have as their genesis the Ellman Lectures I delivered at Emory University in Atlanta, Georgia, in the fall of 2010. The first chapter, "Flying Rabbits," explores my involvement with SF and superheroes as a child, with some thoughts on the deep origins of such superhero features as body-hugging outfits, otherworldly origins, double identities, and flying. The second chapter, "Burning Bushes," is concerned with my undergraduate interest in ancient mythologies, which both pre-date and inform SF. It then goes on to speculate on the differences between realistic fictions and the other kinds, and on the positive and negative capabilities of each.

The third chapter, "Dire Cartographies," is partly about my unfinished Ph.D. thesis, which was about a number of nineteenth- and early twentieth-century fictions I collected together under the label "The Metaphysical Romance." What intrigued me about the books I was studying were the supernatural female figures in them, the realms these inhabited, and the Wordsworthian/ Darwinian split in the visions of nature they represented. These explorations led me to utopias and dystopias, of which the Victorian non-realists and those who continued to write in their traditions were very fond. "Dire Cartographies" is thus also about the three novels I myself have so far written that might be viewed as a continuation of these literary traditions.

The second part, "Other Deliberations," gathers together some of my many pieces of writing about specific works of SF over the years. Some are reviews, some are introductions, others were originally radio talks. Why did I choose these particular works of SF to write about? you may wonder. But I didn't choose them, exactly: in each case, someone else asked me to write about them, and I was unable to resist.

The third part, "Five Tributes," is a group of my own mini-SF pieces. These were selected from among the many such that I have written over the decades, and each draws on recognizable memes from the genre. Four are self-contained miniatures, but the last one—"The Peach Women of Aa'A"—is from my novel *The Blind Assassin,* one of whose main characters is a writer of pulp-magazine science fiction during the early years of what is referred to as the golden age.

So that is what this book is about. It's about my somewhat tangled personal history with SF, first as a child, then as an adolescent, then as a one-time student and academic, then as a reviewer and commentator, and then, finally, as a composer.

.....

But where does all of this come from—the reading, the writing, the engagement, and especially the wilder storms on the wilder seas of invention? Everyone wants to know this about writers: *What is your inspiration, what put you up to it?* They're never satisfied with such explanations as "Because it was there" or "I don't know what came over me." They want specifics.

So let me try this:

As a young child, living briefly in the winter of 1944–5 in an old house in Sault Ste. Marie, I used to get up before anyone else was awake and climb to the cold but spacious attic, where in a state of solipsistic bliss I would build strange habitations and quasi-people with a bunch of sticks and spools called Tinkertoy. What I really wanted to make was the windmill pictured on the box, but my set didn't have the necessary parts, and as it was wartime I was unlikely ever to possess the missing items.

Some say that the art one makes as an adult supplies the absence of things longed for in childhood. I don't know whether or not this is true. If I'd been able to create that windmill, would I have become a writer? Would I have become a writer of SF? We'll never know the answer to that question, but it's one theory.

Meanwhile—in gravely altered form—here is the windmill. I hope you have as much fun with it as I have had.

NOTES

1 The quotation by Octavia Butler appears in the About the Author note at the back of her novel *Parable of the Sower.*

1 *Dick and Jane* was a school reader series of the 1940s.

2 The *New Scientist* article appeared in the November 18, 2008, issue, under the general heading "The future of a genre."

4 "The Wall" is the Berlin Wall.

4 The Lizard Men of Xenor appear in my novel *The Blind Assassin*, in the chapter of that name.

5 Ursula K. Le Guin's review appears in the *Guardian*, August 29, 2009.

6 The public discussion with Ursula Le Guin took place in Portland, Oregon, on September 23, 2010, as part of the Portland Arts and Lectures series.

7 Bruce Sterling's essay "Slipstream" was originally published in *SF Eye #5*, July 1989.

8 *Des Knaben Wunderhorn* was a collection of German folkloric material published between 1805 and 1808.

11 Tinkertoy was a pre-Lego assembly set.

In Other Worlds

SF and the Human Imagination

Flying Rabbits: Denizens of Distant Spaces

The child was already in the air, buoyed on his wings, which he did not flap to and fro as a bird does, but which were elevated over his head, and seemed to bear him steadily aloft without effort of his own.

EDWARD BULWER-LYTTON, *The Coming Race*

I have spoken of the shaman and the folktale hero, of privation that is transformed into lightness and makes possible a flight into a realm where every need is magically fulfilled.

ITALO CALVINO, "Lightness," *Six Memos for the New Millennium*

That which we do not bring to consciousness appears in our lives as fate.

C. G. JUNG

I entered the sort of modern wonder-tale world we might generally label SF at an early age. I grew up largely in the north woods of Canada, where our family spent the springs, summers, and falls. My access to cultural institutions and artifacts was limited: not only were there no electrical appliances, furnaces, flush toilets, schools, or grocery stores, there was no TV, no radio shows available except for those on short-wave Russian stations, no movies, no theatre, and no libraries. But there were a lot of books. These ranged from scientific textbooks to detective novels, with everything in between. I was never told I couldn't read any of them, however unsuitable some of them may have been.

I learned to read early so I could read the comic strips because nobody else would take the time to read them out loud to me. The newspaper comics pages were called, then, the funny papers, although a lot of the strips were not funny but highly dramatic, like *Terry and the Pirates*, which featured a femme fatale called "The Dragon Lady" who used an amazingly long cigarette holder, or oddly surreal, like *Little Orphan Annie*—where were her eyes? The funny papers raised many questions in my young mind, some of which remain unanswered to this day. What exactly happened when Mandrake the Magician "gestured hypnotically"? Why did the Princess Snowflower character go around with a cauliflower on either ear? And if those weren't cauliflowers, what were they?

In addition to being a comics reader, I was an early writer, and I drew a lot: drawing and reading were the main recreations available in the woods, especially when it was raining. Very little of what I wrote or drew was in any way naturalistic, and in this I suspect I was like other children. Those under the age of eight gravitate more easily toward talking animals, dinosaurs, giants, flying humanoids of one kind or another—whether fairies, angels, or aliens—than they do to, say, portrayals of cozy domestic interiors or bucolic landscapes. "Draw a flower" was what we used to be taught in school, and by that was meant a tulip or a daffodil. But the kinds of flowers we really liked to draw had more in common with Venus flytraps, only a lot bigger, and with half-digested arms and legs sticking out of them.

I revisited my early non-naturalistic tendencies during a

recent trip I took through my own juvenilia, or what survives of it. When I say "juvenilia," I'm not talking about the precocious teenage poems of William Blake or John Keats, but about things I was doing in the mid-1940s when I was six or seven. They centred around my superheroes, who were flying rabbits. Their names were Blue Bunny and White Bunny, and they were modelled upon two unimaginatively named real-life stuffed animals who did indeed go flying through the air, propelled by an age-old technology called "throwing." But it wasn't long before these feeble heroes morphed into two tougher creatures called Steel Bunny and Dotty Bunny, who flew in a more conventional superhero way, by means of capes. Steel's cape had bars on it, Dotty's had dots. So far, so clear.

My superhero rabbits were pale imitations of my older brother's more richly endowed creations. It was he who invented flying rabbits—extraterrestrial flying rabbits. His were equipped with vehicles and advanced technologies—spaceships, airplanes, weaponry, the lot—and did battle not only with their hereditary enemies, the evil foxes, but with robots and man-eating plants and lethal animals. The planet where my brother's rabbits lived was called Bunnyland; mine inhabited a more mysterious place called Mischiefland. Now what impelled me to name it that?

The rabbits in Mischiefland led a disorganized existence. They floated around by means of balloons—unavailable during the Second World War and thus of great fascination to me. Also I had by this time read *The Wonderful Wizard of Oz,* in which the wizard goes soaring away in a basket lifted by an enormous hot-air balloon. I allowed not only my rabbits but their pet cats to be levitated in this way. (I was not permitted to have a cat, and longed for one, so my rabbits had a lot of them.) The rabbits ate nothing but ice-cream cones, rare and desirable during wartime and the several lean years that followed. And they did tricks: specifically, a lot of twirling in the air, with the aid of their flying capes. They were only fitfully interested in shooting guns, pursuing criminals, saving the world, and so forth, though they did eject the occasional bullet from the occasional handgun, smiling eerily while doing it. But mostly, it seems, they just wanted to have fun and fool people.

Where did we kids discover the knowledge of flying capes, superpowers, other planets, and the like? In part, through the primitive comic-strip superheroes of the times, the most popular of which were Flash Gordon, for space travel and robots; Superman and Captain Marvel, for extra strength, superpowers, and cape-based flying; and Batman, who was a mortal, with a nonfunctional cape—one that must have encumbered him somewhat as he clawed his way up the sides of buildings—but who nonetheless shared with Captain Marvel and Superman a weak or fatuous second identity that acted as a disguise. (Captain Marvel was Billy Batson, the crippled newsboy; Superman was Clark Kent, the bespectacled reporter; Batman was Bruce Wayne, the very rich playboy who lounged around in a smoking jacket.)

Those—crossed with *The Wonderful Wizard of Oz,* a dollop of Greek mythology, and our one small book on the solar system—were probably the sources of our core ideas. The solar-system book was sedate in and of itself, but I should point out that at that time the planets were still relatively unknown, and thus still open for population by extraterrestrial life, which in our case leaned heavily toward hostile humanoid aliens with one eye and three-fingered hands, animals with razor-sharp teeth and nasty lurking and disembowelling habits, fish that could shoot electric rays at you or gas you to death, and plant life equipped with poisonous prickles or bulbs, or whiplike tentacles and rapid digestive systems. As our father was an entomologist and all-round naturalist, we also had ample access to scientific drawings of, for instance, pond life under the microscope, which may have contributed to our ideas of what Martians and Venusians and Neptunians and Saturnians should look like.

As for disguises, I note that our rabbits seldom felt the need for them: being short and young, we were our own Billy Batsons, and I assume that projecting your child ego onto a flying rabbit was enough of a dédoublement for us.

But where did the creators of the superheroes in the funny papers get their *own* ideas? I now find myself wondering. *Ex nihilo nihil fit*: from what ur-stock did these early superheroes descend? Evidently there were some key gene pools: Superman came from the Planet Krypton, so was clearly a child—in part—of the science fiction of the 1930s, which was filled with the letters *K* and *Z* and *Y* and *X* and *Q*—those oddities of the alphabet.

Captain Marvel's magic word, SHAZAM, was composed of the initial letters of a number of classical gods and one non-classical figure—Solomon, Hercules, Atlas, Zeus, Achilles, and Mercury—so he descends to us, in part, through ancient mythology. Indeed, Captain Marvel's mentor, the wizard Shazamo, once palled around with the enchantress Circe, she of the man-to-pig transformational powers in *The Odyssey*. I think the creators of the Big Red Cheese must have read the same books that I myself read as a child. (Wonder Woman also sports this line of descent, with her links to the goddess Diana the Huntress, she of the chastity and silver bow, the bowstring of which must have become—we just know it!—Wonder Woman's potent lasso. In her early life—

that is, in the comic books of the 1940s—Diana Prince, Wonder Woman's alter ego, turns to jelly and loses power whenever kissed by her love object, Steve Trevor; virginity being an attribute of the original goddess.)

Batman, on the other hand, is born of technology alone. He is entirely human and therefore touchingly mortal, but he does have a lot of bat-machinery and bat-gizmos to help him in his fight against crime. The contemporary magazine most pertinent to him would thus not have been *Weird Tales* but *Popular Mechanics.* He is also—from the point of view of style and decor—the most futuristic of the superheroes: Gotham City, in its first iterations, was highly streamlined, with pronounced art deco influences.

Mythology, science fiction of the other-planetary kind, and modern technology: they all do fit together. At first glimpse, mythology might seem to be the odd one out, being ancient rather than ultra-modern; but as we have seen in the cases of Wonder Woman and Captain Marvel, this is very far from being the case.

In fact, all of the most salient features of these early comic-book heroes—and thus of my own flying rabbits, closely related to them except for the ears and tails—have deep roots in literary and cultural history, and possibly in the human psyche itself.

OTHER WORLDS

Where do other worlds and alien beings come from? Why do young children so routinely fear that there is something horrible under the bed, other than their slippers? Is the under-bed monster an archetype left over from prehistory, when we were hunted by cave tigers, or is it something else? Why do young children also believe that such inanimate objects as spoons and stones—let alone their stuffed teddy bears—have thoughts like theirs, and good and bad intentions toward them? Are these three questions related?

The ability to see things from the point of view of another being has been receiving a lot of attention from biologists lately, most notably Frans de Waal in his book *The Age of Empathy*. It used to be thought that only human beings could imagine life from the position of another, but not so, it seems. Elephants can, and chimpanzees, but not monkeys. Only a being with a sense of "self" can do this, it is supposed. One way of testing for this sense-of-self capacity is through mirrors. Does an animal looking at its reflection in a mirror recognize the reflection as itself? Intriguing experiments have been done in which elephants are presented with elephant-sized mirrors, having first had a visible mark painted on one side of the head and an invisible mark painted on the other side to exclude the sense of touch as a factor. If the elephant sees the mark on its reflected image and then touches the real mark on its head with its trunk, it must know that the reflection is "itself." Often, before coming to the realization that the reflection is indeed itself, an elephant will look behind the mirror. So will a human child.

If you can image—or imagine—yourself, you can image—or imagine—a being not-yourself; and you can also imagine how such a being may see the world, a world that includes you. You can see yourself from outside. To the imagined being, you may look like a cherished loved one or a potential friend, or you may look like a tasty dinner or a bitter enemy. When a young child is imagining what's under the bed, it is also imagining what it might represent to that unseen creature: usually prey. It is possibly not a good idea to tell the little ones that they look good enough to eat. Frisky the Cat wouldn't be bothered by such a statement, lacking as she does a capacity for empathy, but Charlie the Child may well have hysterics.

One of the more brilliant innovations of H. G. Wells's *The War of the Worlds* is that it so clearly sets forth what we puny human beings might look like to godlike intellects far superior to ours. From that time to this, we've been told many stories along these lines. Or, as Shakespeare put it, about gods thought of as somewhat closer to home than Mars: "As flies to wanton boys are we to th' gods, / They kill us for their sport."

.....

Other worlds with strange inhabitants have been numerous in human mythologies and literatures. I'd speculate that, including all the fantasylands devised by children that never see publication, there are many more imaginary locations than there are real ones. Whether they are places we go after death—good or bad—or homes of the gods or supernaturals, or lost civilizations, or planets in a galaxy far, far away, they all have this in common: they aren't here and now. They may be long ago or far away; they may be situated in that nebulous region, "the future"; they may even be given real estate in "another dimension" of the space-time we ourselves inhabit. The convention seems to be that other beings can pop into our living rooms from somewhere else, but they can't drag along the entire other world from which they come. We, on the other hand, can slip through a cupboard or through a wormhole in space and find ourselves transported to their realm. Stories about encounters with other beings thus always involve travel, one way or another. Something or someone moves from "there" to "here," or we ourselves move from "here" to "there." Portals, gateways, waystations, and vehicles abound, as in—come to think of it—ancient myths, with their cave entrances and chariots of fire.

Our ability to conceive of imaginary places—a somewhere that isn't immediately tangible in the way that the dinnertime pork chop is tangible—appears very early in our individual lives. At first—when we're extremely young—it's a case of out of sight, out of mind: objects hidden from our view simply disappear, then appear again. It takes us a while to figure out that the rubber duck that went behind the curtain is still somewhere rather than nowhere.

Once we've decided things go to another place rather than simply ceasing to exist, we find it hard to shake that notion. Being "here," then suddenly not being here: is that where the concepts of, for instance, afterlives and teleportation originate? Does *Star Trek*'s Scotty derive his ability to beam people up from the discovery that the rubber duck in our early games of peek-a-boo was still there really? Is dead Granddad floating around in the spirit world trying to get in touch with us? And will we, too, float around in that way, since it is very hard to picture the self as being nowhere at all? Surely the dead go somewhere, other than the tomb. Once, they went to the Egyptian Afterlife to get their souls weighed, or to the Fields of Asphodel, or up into the sky to become constellations, or to a physical location called Heaven. Now, perhaps, they might go to the Planet Krypton or wherever it is that E.T. went. And are the Fields of Asphodel and the Planet Krypton more or less the same place?

One method of approaching Other Worlds would be to trace

their literary lines of descent—from the Mesopotamian underworld to the Egyptian Afterlife to the Domain of Pluto to the Christian Hell and Heaven to the Utopia of Sir Thomas More to the Islands of the Houyhnhnms and Dr. Moreau, and finally to Planet X and Gethen and Chiron. But Other Worlds have existed in many cultures, within which they can trace many separate literary and cultural lines of descent. Could it be that the tendency to produce such worlds is an essential property of the human imagination, via the limbic system and the neocortex, just as empathy is?

THE OUTFITS

Once upon a time, superhuman beings wore robes, like angels, or nothing, like devils, but the twentieth-century superhero outfit has more proximate fashion origins. The skin-tight clothing with the bathing suit over the abdominal parts, the wide, fancy belt, and the calf-high boots most probably derive from archaic turn-of-the century circus attire, especially that of high-wire artists and strongmen. (With pleasing circularity, the stars of World Wide Wrestling now dress up in costumes similar to those of comic-book characters whose own colourful and six-pack-disclosing attire recalls that of earlier bemuscled showmen.)

The cape may descend from the knights so prominent in the Pre-Raphaelite art that would have been familiar to the originators of these figures, or—closer to hand—from stage magicians, or, at a stretch, from Bela Lugosi as Dracula, in the black-and-white film of that name, back when vampires were vile and smelled bad rather than being sparkly in the sunlight and love's young dream, as they seem to have become today. There was also the Cloak of Invisibility that featured in old folk tales, resurfaced in modern scientific-discovery dress in Wells's *Invisible Man,* made a reappearance in its original magic form in the Harry Potter books, and became a new kind of camouflage material in William Gibson's *Neuromancer.* But none of the early 1940s comic-book superheroes had a Cloak of Invisibility, probably because it was hard to draw a picture of a person being invisible. (The closest we come is perhaps Wonder Woman's transparent helicopter, indicated by a dotted line.)

The mask was not obligatory for superheroes: neither Superman nor Captain Marvel needed such an identity-concealer, as each had a whole other body to slip into. (Clark Kent's ability to peel off his reporter suit in a phone booth and suddenly expand into someone a great deal bigger and more muscular, like one of those dried-gel Santa Clauses you drop into water, was never adequately explained.) Batman's mask may have come from the *commedia dell'arte* tradition, or from knights-incognito such as Ivanhoe. Or—and these are more sinister origins—from the Phantom of the Opera, or from Fantômas, a masked and also French evil genius from the turn of the century. Or possibly just from the standard masked robber of the comics. As Batman himself was mortal and did not transform from one bodily shape into another, you can see why he would need a mask.

Outfits—or special costumes and regalia—are of course very old. We are familiar with ceremonies such as university graduations—you are presented with a hood item or hat or scroll, and thus become something you weren't before. At the investiture ceremonies of popes, the new pope is given the Fisherman's Ring, the wearing of which grants him, in the eyes of believing others, a huge amount of spiritual power that the individual man would not have without this symbol. (Rings have had special abilities for a long time; see the magic rings in *One Thousand and One Nights*, as well as Richard Wagner's *Ring Cycle* and J. R. R. Tolkien's not unrelated *The Lord of the Rings*, both of which draw on much earlier traditions.) In coronations, the magic items are the crown and the sceptre: they embody the role of king, as the king was once thought to embody the realm he ruled. The further back you go, the more significant the objects worn or carried become. In the age of the god-kings, such as those of ancient Egypt or Sumer, the man or woman and the costume and regalia were almost one and the same: you were the role, and the role was the garment and its embellishments. You inhabited it rather than just wearing it.

Consider the oldest poem we know anything about, the Mesopotamian cycle sometimes called "Inanna's Journey to Hell." In it, the life-goddess Inanna descends to the Underworld to confront her sister, the goddess of death, Ereshkigal. To defend herself during the trip, Inanna puts on and carries an astonishing number of talismanic and powerful objects: the special sandals, the seven insignia, the desert crown, a queenly wig, a rod, a number of gems, two pectorals, a golden ring, some face makeup, and a robe of sovereignty. But the laws of the Underworld say that she must give up each one—you couldn't take it with you, even then—and when all her protective charms are gone, she's naked; whereupon she dies and is hung up on a spike. For every Achilles there's a heel, a condition of vulnerability; for every Superman there's a kryptonite, a force that negates special powers.

The Mesopotamian story does have a somewhat happy outcome. Inanna is the goddess of life and reproduction, so it would be a catastrophe for her to remain in the land of the dead. But no mortal can be sent to the Underworld to resurrect her with the Water of Life, since any mortal who goes there will die; so the god Enkil makes two non-human beings from the dirt beneath his fingernails and sends them down instead, thus giving us—we might say—the ancestors of Golems, and statues that come to life, and, ultimately, robots. We are not told that on her journey back to the upper world Inanna regains all her regalia, but it must have been so because later in the poem she is again wearing her crown of authority.

How much older than Mesopotamia is the connection between special clothing and talismans and heightened powers? Quite a lot older. Some of the very few human figures in Paleolithic cave paintings are in fact semi-human: they are thought to be shamans who by putting on the skins and horns of animals become part animal themselves, and thus able to join the animals in thought, to determine their whereabouts, and perhaps to ask them to make a gift of their bodies to the hungry tribe.

It's the outfit and the ritual associated with it that embodies the shamanistic power. The shamans of hunter-gatherers lived with the community, not in a palace or temple. Most of the time they went about their daily lives like everyone else, but when occasion required it they transformed themselves into their magic alter egos in order to serve the community. There's an Australian Aboriginal film called *Ten Canoes*, set in pre-contact days, in which we can see this transformation taking place. The shaman's powers are needed; he steps behind some bushes and emerges in full body paint, ready for magic. He is two people: his ordinary self and his other self, powerful in extraordinary ways, and able to travel between the seen and the unseen. His special decoration, just like Captain Marvel's, is a signal to the watchers that he is in his altered state.

THE DOUBLE IDENTITY

The doubleness of superheroes thus has a very long ancestry. But more immediate ancestors abound in the period just preceding the advent of the comic book.

In nineteenth-century fiction, doubles are plentiful, as they are, indeed, in nineteenth-century opera and ballet—think of the white and black swan princesses of *Swan Lake.* Stevenson's Dr. Jekyll and his shorter, younger, and hairier double, the nasty Mr. Hyde, Wilde's Dorian Gray and his diseased and corrupt Picture, and Poe's William Wilson and his taunting twin are among the best-known literary examples. Some speculate that such good/bad pairs may have at least some roots in the lives of real people—such as Jonathan Wild the thief-taker, whose secret life was as a mastermind of crime, or Deacon Brodie of Edinburgh, a respectable gentleman whose midnight misdeeds are thought to have inspired Stevenson.

But these are sinister doubles. For the weak or frivolous alter ego acting as a front for the strong, virtuous hero—more like Clark Kent and Superman—we should most likely be looking at the Scarlet Pimpernel—dithering fop by day, steel-nerved rescuer by night—and possibly even Alexandre Dumas's *The Count of Monte Cristo,* in which the hero assumes several aliases—including that of an eccentric English lord—in order to reward virtue and punish crime. Sherlock Holmes, that supreme intellect, clue-tracker, and criminal-hunter, was a master of disguise, often posing as someone less than he really was: a feeble, kindly old clergyman, for instance, or an out-of-work groom.

In addition to his disguising "normal" alter ego, the superhero of the 1940s was required to have a powerful enemy or two. Carl Jung made no secret of the fact that he based much of his mapping of the psyche on literature and art. For instance, his theory of the "Shadow"—that dark double of the Self—has a great deal in common with, for instance, *The Tales of Hoffmann*, or indeed any of the "double" narratives I've already mentioned. A comic-book character leading a split life and engaged in a battle between Good and Evil might well be expected to show Jungian characteristics, and in fact Batman is an almost perfect case study.

Batman has three main enemies, who to a Jungian would obviously be projections of Bruce Wayne that Wayne himself has not come to terms with. (In Blakean terms, the two male enemies would be called his Spectres and the female one might be his Emanation.) For Bruce, the female element is conflicted—he's a confirmed bachelor, and has no nice-girl Lois Lane sentimental figure in his life. But the sinuous and desirable Catwoman with whom he frequently skirmishes must be his Jungian "dark anima" figure: even a child could recognize that there was a lot of unresolved electricity going on between those two.

The sadistic card-playing Joker, with his sinister-clown appearance, is Batman's Jungian Shadow—his own interest in dress-ups and jokes turned malicious. There's another Shadow villain—the Penguin—who wears an outfit reminiscent of period cartoons of capitalists, with spats, cigarette holder, and top hat. His civilian alias even has a three-barrelled, pretentious, old-plutocrat faux-English name: Oswald Chesterfield Cobblepot. The Penguin is the "rich" side of playboy Bruce Wayne gone rancid.

Then there's Robin, the Boy Wonder, who is Bruce's ward. Is Bruce gay? Don't even think about it. From the point of view of we mythosophists, Robin is an elemental spirit, like Shakespeare's Puck and Ariel—note the bird name, which links him to air. His function in the plot is to aid the benevolent master trickster, Batman, with his plans. From the point of view of we Jungians, however, Robin is a Peter Pan figure—he never grows up—and he represents the repressed child within Bruce Wayne, whose parents, you'll recall, were murdered when he was very young, thus stunting Bruce's emotional growth.

This is the kind of hay, or perhaps hash, that can be made of such comic-book superheroes once you really get going. Both they and Jung himself can be viewed through Hoffmannesque magic spectacles and seen to be part of the same mythology.

But from the point of view of we kids—the primary readers—Robin was simply ourselves—what we would be if we, too, had masks and capes and could go running around in them under the delusion that nobody would know who we were, and—better still—stay up long after our bedtimes, allowed to participate in the doings of what we fondly hoped was the adult world.

THE FLYING

Batman couldn't really fly. This must have dampened my view of him somewhat, since—judging from the pictorial evidence—flying was the superhero attribute that interested me the most during my days as a child superhero cartoonist. Almost everything in my created world of Mischiefland was airborne. Why was I so keen on the life of the air? Come to think of it, why were the creators of the superheroes so keen on it as well?

The interest appears to be widespread. One of the recent—though minor—superheroes I've come across is named Kidney Boy. I picked him up on the Internet micro-blogging site Twitter, and, intrigued by his nom de plume, I offered to design a superhero outfit for him, complete with special powers and charm-word. In real life, Kidney Boy has a somewhat geeky alter ego—he's a nephrologist, or kidney doctor. He told me he would love to have a magic power, one that would allow him to create new kidneys that would be perfect matches for his dialysis patients. But if he couldn't have that, he said, could he please have "the flying-around thing"?

In the event, I provided everything he wanted: an outfit with a purple kidney helmet; a magic scalpel that would never fail; a magic word—Nephro-Change-O!—that would not only create the desired kidneys but cause them to slide effortlessly into his patients without even an incision; and, to top it all off, "the flying-around thing."

Ontogeny repeats phylogeny—could Kidney Boy and I both have inherited this interest in flying? Is it inscribed in our genes, or is it the result of a meme, of the kind popularized by Richard Dawkins—a theme, idea, or motif passed down from generation to generation, self-replicating, mutating, and competing with other memes as it goes? In either case, it's surely no accident that the ability to fly, with or without the aid of wings, flying shoes or capes or horses or carpets, balloons, aerodynamic kidneys, and so forth, has a very long history.

What does the ability to fly portend, in a superhero or even a god? We are not talking here about airplanes and helicopters: the flying-around thing is not about more rapid and efficient real-life transportation methods. It has to do with wings, either actual or implied, with rising above the earth, and with the ability to glide effortlessly from one place to another. It has to do with overcoming the restrictions of the body, that dead weight of ultimate mortality we lug around with us. "If I had the wings of an angel," mourns the old folk song, "over these prison walls I would fly . . ." We don't have them. But it seems we've always wanted them.

At first glance, you might think that wings are an unconditionally good thing. In fact, wing-possession in a non-human being is an alarm signal.

For instance: Inanna, the life-and-sex goddess of Mesopotamia already mentioned, was shown with wings, but she was definitely an entity you wouldn't want to get tangled up with. Both she and her later incarnation, Ishtar—who appears in *The Epic of Gilgamesh*—were travellers between realms—Earth to the Underworld, Earth to Heaven—and both were known for seducing mortal men who then met tragic fates. When Ishtar asks Gilgamesh to be her husband, he recites a long list of her former lovers whom she has killed or tortured or turned into wolves or dwarfs.

The Greeks had two messenger gods: Iris, a morally neutral figure who had golden wings, and Hermes, who governed communications (thus justifying his handsome, curly-headed appearance on the Bell telephone books of the 1940s, with his traditional winged helmet and sandals, but with a modern touch—a number of thick telephone cables wrapped modestly around his midsection). Hermes was also the god of travellers and as such conducted the souls of the dead to the Underworld, so going with him on a journey was not always a fortunate thing to do. There was also

Nike, thought of as the Goddess of Victory, but perhaps more accurately the goddess who announces victory—another messenger. She, too, had wings. But victory for one side, as we know, always means defeat for the other.

In the Judeo-Christian tradition, messengers from the divine realm are called "angels," which is simply the Greek word for "messenger." The Hebrew word has the same meaning. In the Bible, angels are not often described as having wings; more usually they appear as human beings, though the seraphim in Isaiah:6 have six wings and some New Testament angels evidently have the power of flight and/or instant teleportation. For their appearances in later art, the two-winged angel images most likely filched their wings from Nike or Iris, or, for the young cherubs, from Eros, the boy-god of love. But whether wingless or not, angels certainly illustrate the troubling nature of messengers. How much fun is it to be told that your hometown is about to be destroyed by fire and brimstone, or that you, an unmarried virgin, are about to get pregnant? The expressions on the faces of Renaissance Virgin Marys are usually ones of apprehension, not joy. A visit from Iris or Hermes or any Judeo-Christian angel messenger was as likely to be bad news as good.

Thus the fact that such divine beings can fly should not necessarily make us trust them. Like the utterings of oracles, the messages they bring are often deeply ambiguous.

TRANSFORMATIONS AND TRICKS

Hermes, the wing-enabled flying messenger, is not only the god of communication, he is also the god of thieves, lies, and jokes.

That's another interesting thing about many airborne non-humans—their odd sense of humour, the delight they seem to take in misleading human beings and playing tricks on them. In the plays of Shakespeare, there are, as I've mentioned, two notable non-human flying beings: Puck in *A Midsummer Night's Dream* and Ariel in *The Tempest.* Both are messengers and servants, carrying out the plans and delivering the decrees of Oberon and Prospero, respectively; and both are disguise artists and trick-players. Do they have their origin in winged Eros (or Cupid), the notoriously practical-joking boy-god of love, messenger of the goddess Venus? Cupid may bring boxes of chocolates today, but formerly he shot his wounding arrows of desire into people who then went crazy with lust and obsessive longing while he himself laughed. The djinni of *One Thousand and One Nights* tales are similar messenger-servants, as are the winged monkeys in *The Wonderful Wizard of Oz*: airborne, powerful, hard to control except through magic. The morally dubious fairies of English folklore bear a family resemblance: disguising themselves and fooling people are what they seem to take most pride in. Puck is strongly of this lineage, thinking it great fun to turn himself into a stool and then whisk out of the way just as someone is about to sit down on him. Making fools of already foolish mortals is his main game.

Come to think of it, such predilections are shared by the early comic-book superheroes. They weren't as a rule crude joke-players, but their transformations certainly involved deception—no one was supposed to know that Clark Kent was really Superman, and vice versa. The episodes of most interest to us child readers were not the maiden-rescuing or the parts where Gotham City gets saved from destruction, or even the hand-to-hand sock-bam-pow battles with villains, but the moments of transformation. First, the bespectacled weakling or crippled newsboy, with all the humiliation implied by that role; then off came the disguise, and the true, strong hero sprang into view like a husband shooting out of a closet in a Feydeau farce—surprise!—and the bad folks quailed, and the bullies could no longer kick sand in your face at the beach. It was the notion of deceiving people that we really liked—the idea that you could walk around among unsuspecting adults—the people on the street in the comic books—knowing something about yourself that they didn't know: that you secretly had the power to astonish them.

In this respect, the 1940s superhero Plastic Man was the champion. His superpower was stretchiness. Because he was plastic—he'd suffered from an unpleasant encounter with a vat of chemicals, which is the modern-day equivalent of having a god for a parent, or being dipped into the River Styx like Achilles—he could mould himself into any everyday object, such as a lamp or ashtray; eavesdrop on everything the crooks and thugs were planning; then leap forth, manifest himself, and wrap his entire body around the evildoers like a long rubber band. He was probably the trickiest, wittiest, and least violent of all the superheroes; more Puck than Oberon, a sort of comic party toy.

The fascination with disguise is very ancient. The gods frequently assumed mortal shapes, the better to walk among humankind unobserved. (This habit was taken up later by folk-tale sultans and kings and even saints, most notably Saint Peter.) The first self-consciously disguised character we meet in literature—or the first who isn't a god, to my knowledge—is the wily Odysseus of *The Odyssey*, who, having been away from home for many years, dresses himself as a ragged beggar upon his return to his palace, where a large number of insolent young men are eating up his herd animals, raping his maids, and trying to marry his wife. Imagine their astonishment when he strings his own superbow—the magic weapon no one else can handle—steps back into his role of king, and kills the whole lot of them. The two gods who take a special interest in Odysseus are Athene, who values intellect and quick wits, and our old friend Hermes, the trickster god of artifice and jokes.

Which brings us back to the flying rabbits I was drawing and telling stories about at the age of six or seven. Now we understood why the planet they inhabited was called Mischiefland; though, back then, I myself didn't know why I'd given it that name. Like many artists, I did it because, well, it just seemed right to me at the time. Balloons, flying, superpowers, mischief: they all went together. Though my superheroes were probably the only ones that had long floppy ears and fluffy white tails.

NOTES

15 Bulwer-Lytton's *The Coming Race* (1871) presents a race of superior human beings who live in a vast cavern underground, harnessing an inner, electrical life force called *vril* for power. (Vril gave its name to the beef tea "Bovril"; bovine vril.) The Vril-ya fly around on vril-powered wings and display superintelligence; the women among them are bigger and stronger than the men, whom they have to treat well lest the latter fly away.

15 The quotation from C. G. Jung is taken from *Meeting the Shadow: The Hidden Power of the Dark Side of Human Nature* (Abrams and Zweig, 1991).

16 Mandrake the Magician is said to have been the first comic superhero, but his hypnotic gesturing was anticipated by Dr. Caligari and Dr. Mabuse, the villains of the two eponymous films that feature their wicked hypnotizing powers.

16 Princess Snowflower was found in the comic strip *Steve Canyon.*

17 The juvenilia is now in the Fisher Library at the University of Toronto.

17 L. Frank Baum, *The Wonderful Wizard of Oz*, 1900.

19 "The Big Red Cheese" is a nickname for Captain Marvel.

19 Diana the Huntress, Artemis: Roman and Greek moon goddesses characterized by virginity, bowmanship, and an affinity with wild animals.

21 Frans de Waal, *The Age of Empathy: Nature's Lesson for a Kinder Society*, 2010.

22 The Shakespeare quotation is from *King Lear,* IV, 1, 32–34.

23 *Star Trek*: a long-running space serial.

23 The Fields of Asphodel were in the Greek Underworld. The Planet Krypton was Superman's home planet.

24 Sparkly vampires can be found in the *Twilight* series by Stephenie Meyer.

25 The Cloak of Invisibility is a feature in folklore; see *Grimm's Fairy Tales.*

28 I first encountered the duplicity of Jonathan Wild in the 1840 novel *Jack Sheppard* by Harrison Ainsworth.

28 The Scarlet Pimpernel is the hero of the 1903 play and subsequent novel by Baroness Orczy.

29 *The Tales of Hoffmann*: In the 1881 opera by Offenbach, all of the villains are traditionally played by the same singer.

33 Nike: Does this swift-flying goddess determine victory or simply reward it? Accounts differ, but either way it's a good name for running shoes.

35 Georges Feydeau wrote many farces that depend on impeccable entry-and-exit timing.

Burning Bushes: Why Heaven and Hell Went to Planet X

Those finds concern religious beliefs prevalent during the Eighth Dynasty of Ammer-Ka; they speak of various Perils—Black, Red, Yellow—evidently cabalistic incantations connected in some way with the mysterious deity Rayss, to whom burnt offerings were apparently made.

STANISLAW LEM, *Memoirs Found in a Bathtub*

[Science fiction is] a mode of romance with a strong inherent tendency towards myth.

NORTHROP FRYE, *Anatomy of Criticism*

My superhero-creating or flying-rabbit phase ended when I was eight. By the time I was nine or ten, I had become a confirmed under-the-covers midnight flashlight reader, devoting myself not only to adventure stories but also to comic books of an increasingly wide variety. In my daytime life, I would read anything that was handy, including cereal boxes, washroom graffiti, *Reader's Digest*s, magazine advertisements, rainy-day hobby books, billboards, and trashy pulps. From this you might conclude that I quite possibly have never been an entirely serious-minded person, or perhaps that I simply have eclectic tastes and like to rummage. Given a choice between a stroll in a classic eighteenth-century garden and the chance to paw through someone's junk-filled attic, I would probably choose the attic. Not every time. But often.

As the twig is bent, so the tree grows, they used to say, so I suppose I should reveal what sort of things bent my own twig; for surely at least some of the books that writers eventually produce as adults are precipitated by what they read avidly as children.

Our house had a ready supply of the same kinds of odd and non-naturalist late-Victorian and Edwardian tales that delighted—for instance—Jorge Luis Borges and many of the "magic realists" that emerged from Latin America in the mid-twentieth century. Between the ages of eleven and seventeen, I read M. R. James, the master of the creepy tale, and all of H. G. Wells's fantastical stories—*The War of the Worlds, The Island of Doctor Moreau, The Invisible Man*, "The Country of the Blind," and many more. Arthur Conan Doyle's *The Lost World*, complete with dinosaurs and Primitive Man, was a favourite; so were H. Rider Haggard's once highly popular *King Solomon's Mines, Allan Quatermain*, and *She*, with their lost civilizations frequently ruled over by beautiful, shoulder-baring, drapery-fluttering queens; and whatever derivative *Boy's Own Annual* adventure stories I could get my hands on.

It goes without saying that I was in love with Sherlock Holmes, and, once I got around to it, with Dashiell Hammett's Sam Spade and Raymond Chandler's Philip Marlowe as well. There is something to be said for a greatcoat or trenchcoat, a back alley, and a clenched jaw, and that none of these men *au fond* had much respect for women did not bother me a whit: the blonde usually did it, and I was not a blonde.

I also read a lot of SF. As I proceeded through high school, I dug into the John Wyndhams—*The Day of the Triffids* came out in

1951, *The Midwich Cuckoos* in 1957. I devoured any Ray Bradburys I could get—it was the 1950s, so *The Martian Chronicles* and *Fahrenheit 451* were both available.

I often read this kind of book when I was supposed to be doing my homework. I was, in fact, leading a double life, or even a triple one: the terms *highbrow*, *middlebrow*, and *lowbrow* were much in use at that time—the metaphor was based on some idea of Neanderthals having receding foreheads—but I seemed to have a taste for all three kinds of brow, which I can't say disturbed me. In the classroom we took Shakespeare—a play or two a year—and the romantic and Victorian poets, among others. Over the five years of high school that were mandated there and then, we studied two novels each of George Eliot, Charles Dickens, and Thomas Hardy, and we studied them very thoroughly. So it was serious highbrow business during the day. But after school, I would lower my brow and indulge in my guilty pleasures: *Donovan's Brain*, *The Kraken Wakes*, and their ilk.

This pattern continued once I was at university, except that my escapism expanded to include double-bill B sci-fi movies of the lowest possible brow level. I saw *The Fly* when it first came out, and *The Attack of the 60-Foot Woman*, whose growth in size rendered her semi-transparent, and *The Head That Wouldn't Die*, and *The Creeping Eye*, a giant eye with tentacles that came from Outer Space like a lot of threatening things in those days, and that—when it finally put in an appearance—had tractor treads clearly visible beneath it. Meanwhile, in my highbrow guise, I was making my way through English literature from Anglo-Saxon to T. S. Eliot, and French literature of the eighteenth and nineteenth centuries, and going to Ingmar Bergman and *nouvelle vague* films.

It did not entirely escape my notice that Beowulf and *The*

Creeping Eye had a few things in common. Both had monsters. Both had gore. Both had heroes. Jane Austen's heroines, on the other hand, had money worries instead of talking skulls, and Madame Bovary did not have her still-alive head preserved under a bell jar, complete with Bride of Frankenstein hairdo, but died of over-spending. Could it be that the sensational, monster-ridden tales of the distant past—now sanctified as part of our priceless literary canon—were joined at the hip to the sensational, monster-ridden tales of the present, which were vilified as trash?

Why do people tell or write such wonder-tale stories? Or, more generally: why do they tell or write any stories at all? How did stories originate? What purpose do they serve in our lives? Are they the result of nurture—we learn our stories, as children, from the adults around us—or are they built in, hard-wired into the brain in "template" form, thus causing stories to generate semi-spontaneously if the epigenetic switch for them is turned on?

To speculate further: do stories free the human imagination or tie it up in chains by prescribing "right behaviour," like so many Victorian Christian-pop novels about the virtues of virtuous women? Are narratives a means to enforce social control or a means of escape from it? Is the use of "story" as a synonym for "lie" justified, and if so, are some lies necessary? Are we the slaves of our own stories—our family narratives and dramas, for instance—which compel us to re-enact them? Do stories optimistically help us shape our lives for the better or pessimistically doom us to tragic failure? Do they embody ancient tropes and act out atavistic rituals? Are they essential to us—part of the matrix of our shared humanity? Do we tell them to show off our skills, to unsettle the complacent audience, to flatter rulers, or, as Scheherazade the Queen of Storytelling did, to save our own lives?

Are they the foundational bedrock of our various societies, or possibly even our various nations—whether those nations have been aspiring ones, in the throes of defining themselves; imperial ones, justifying their domination of others; or declining ones, lamenting their own passing? Are they inseparable from our cultures, whether ancient cultures encrusted with age-old symbols or recently formed cultures in search of such mental jewellery?

Or are stories just pastimes—old wives' tales to be spun round a cottage fireplace—or sentimental and sensationalistic novels to be devoured by bored young ladies reclining on nineteenth-century chaises longues, or TV-series spinoffs, a minor part of the entertainment industry—in all cases frivolous by definition? And are slabs of foundational bedrock and frivolous, entertaining pastimes mutually exclusive?

The answers to such questions have varied over the years—indeed, over the millennia—and many heads have literally rolled and many panel discussions have been held over the differences among those answers. It was—once, and in some places—of crucial importance to your continued existence whether you told a story about Augustus Caesar that included divinity among his attributes or whether your story about Abraham had him expelling Hagar and Ishmael into the desert or, on the contrary, claiming Ishmael as his son and co-inheritor. In seventeenth-century New England, your health could be permanently affected by whether your story about witchcraft affirmed or denied its existence. In medieval Europe, it was literally of burning importance to you whether you told a story in which God was three in one, or a story in which he was one alone, or whether your story contained two gods, a good one and a bad one. Orthodox stories of any kind always try to eliminate their competitors.

Stories and the significance given to them can alter very quickly. Five years after the onset of the Salem witchcraft trials and their resulting deaths, several of the New England judges and divines who had egged on the trials were issuing public repentances. "The Devil was indeed among us," said one, "but not in the form we thought." Yesterday's righteously condemned miscreant can become today's martyr, and vice versa. It all depends on the story. But stories themselves, of one sort or another, are always with us, and are always moving and changing through time.

We live in an age of intense speculation about stories and their origins and purposes. Denis Dutton, in his book *The Art Instinct,* proposes the notion that the arts—and also the impulse toward religion—are encoded in our genes. According to this theory, artistic capabilities would of necessity be evolved adaptations, acquired during the roughly two million years the human race spent in the Pleistocene as hunter-gatherers. To have been "selected" in this way, the arts would have had to have conferred some noteworthy benefits on us during those millennia; that is, those who demonstrated such abilities as singing, dancing, the making of images, and—for our purposes—the telling of stories would have had a better chance at survival than those without them. That makes a certain sense: if you could tell your children about the time your grandfather was eaten by a crocodile, right there at the bend in the river, they would be more likely to avoid the same fate. If, that is, they were listening.

During those millennia, so far as we can tell, there was nothing we might now consider a "religion"—a theology with a set of worked-out abstract dogmas and special dwellings set aside for worship, such as temples. Instead, beliefs about the unseen or numinous world were integrated into all life because everything was thought to have a soul or essence. Any action performed would therefore have been deeply significant in a way that we in modern, secular, Westernized society can scarcely imagine. Such a worldview would have been rich and wondrous, but it would also have contained many fears—fear of crossing boundaries, of offending divinities, of breaking taboos. There must have been a very thin line between gods and monsters.

Vestiges of that worldview linger in the early literature that has come down to us. Greek mythology abounds with stories in which people are transformed by gods into natural beings such as animals, birds, and trees; and, in return, such natural entities often speak to or communicate with people. It's noteworthy, too, that in the biblical Book of Exodus God does not appear to Moses in human form; instead he is a voice emanating from the well-known Burning Bush—a bush that is in flames but is not consumed. The bush itself is not God in physical form but an angel or messenger: the narrator of this story is taking care to avoid trapping or limiting God, because a God confined to or circumscribed by a physical object such as a bush—however sacred, inflammable, and loquacious—would be a God that was potentially destructible.

What does the voice say? Among other things, it says that its name is I AM THAT I AM, or possibly I WILL BE WHAT I WILL BE. God is a sort of gerund, a noun-verb, and the burning bush that is not consumed is curiously similar to the vision of the pre-Socratic philosopher Heraclitus, for whom the source of all things was fire—not a physical object but a process.

In any case, hearing a bush speak, especially in such a portentous manner, is not something a Jane Austen heroine would be likely to experience. Such an event might happen in an Ann Radcliffe Gothic shocker, but only if there were a sinister count hidden inside the bush. It could happen easily in a fairy tale, however, or a "fable" like *Alice in Wonderland*, or in a Greek myth at the point at which some hapless maiden is being transformed into a sapling or other item of vegetation by a god. And it could happen effortlessly on Planet X.

. . . . .

All myths are stories, but not all stories are myths: among stories, myths hold a special place.

I went to college in the late 1950s at the University of Toronto—a place and time when thinking about mythology—or very ancient, centrally important stories and their nature and form—was at the top of the agenda. Edmund Carpenter, a noted anthropologist, was co-editing a magazine called *Explorations*, highly influential at the time. Working with him was Marshall McLuhan, soon to become the pre-eminent theorist in the field of media and communications. The first of McLuhan's books was *The Mechanical Bride,* which analyzed such things as advertisements and comic books for their mythic and psychological content, and illustrated itself with reproductions of the ads, and thus got pulled from the market by soap flake companies and such for copyright infringement; but you could buy this book on the sly from McLuhan's basement, which I did. Due to my interest in cereal boxes, magazine ads, and comics, I of course found this book delectable.

A third figure loomed large in the literary world of that time. His name was Northrop Frye, and he taught at Victoria College, which I myself attended. I only took half a course from him—the "Milton" half of "Spenser and Milton"—and he was not a direct influence on the kind of student writing that I was by that time doing; but he validated the literary enterprise itself by underlining its importance to civilizations. He had the added benefit of being a reader at all three brow levels, which pleased me a lot: it's always encouraging to be told that it is intellectually acceptable to read the sorts of things that you like reading anyway.

At that time Frye had published two books of criticism that had taken academia by storm: *Fearful Symmetry*, a study of William Blake's long narrative poems, which Blake called "prophecies," and *Anatomy of Criticism*, a hugely ambitious project that set up a series of templates into which literary works could be slotted according to various overlapping and interlocking characteristics. Myth featured largely in this book.

Every year, Frye gave his famous Bible course, "The Bible as Literature," which was audited by students from miles around. "Are you saying that the Bible is a myth?" one student asked him. "Yes," Frye replied. "That is what I am saying."

But what did he mean by "myth"? His interest in it came from his concern with story forms, and with how literary works and possibly the human imagination itself were structured. For readers, it cleared up some confusions common at the time—confusions about genre and rhetoric, for instance: there was no point in expecting from an apple the same qualities you might find praiseworthy in a steak.

Here's a very simplified version of Frye's theories about how "myth" and its patterns were structured. In Greek mythology there are four ages of man: the golden age, the silver age, the bronze age, and the iron age. These correspond to spring, summer, autumn, and winter, and also—says Frye—to four main types of story: the romance (spring), in which the hero journeys on a quest, kills dragons, and rescues maidens; the comedy (summer), in which the hero and the maiden can't get together due to interference by censorious old fogies, but which, after complications, ends with a marriage; the tragedy (autumn), in which the protagonist falls from a height of power and prestige and ends up dead or in exile; and irony (winter), in which ancient geezers and biddies sit round a fire in a frozen world and tell tales. Strangely enough, the tales told by the geezers usually turn out to be romances, in which the hero journeys on a quest, kills dragons, rescues maidens, and so forth. And so the story cycle begins again.

Set beside this circular, cyclical, seasonal pattern is a linear pattern—one with a beginning moment, a middle journey, and a final endpoint. This pattern is most fully illustrated in Western culture by the Christian Bible, which begins with Creation and ends with Revelation. The Book of Revelation is usually thought of as "the Apocalypse," in which the world comes to an end in awful destruction, four horsemen, plagues, fire, and so on. But it really means what it says: "Apocalypse" means Revelation, the moment when all becomes clear, all is revealed. In the Christian Bible, the Creation is followed by the Fall, which is a tumble from a timeless paradise into a space-time continuum that contains death; or, as Stephen Dedalus says in James Joyce's *Ulysses*: "History is a nightmare from which I am trying to awake."

In between the bookends of Creation and Revelation are a large number of stories, culminating in the life and death and reappearance and ascension of Christ, and finally the Apocalypse, the Last Judgment, and the arrival of a perfect Heavenly society, called the New Jerusalem—as opposed to the bad society also presented, usually identified with Babylon and/or Rome. The story ends there rather than beginning again—as it does in cyclical myth structures—since presumably once a thing is perfect and complete, nothing can change. "Forever and ever" means that time has come to an end.

Frye also explores a set of archetypes, each of which has a positive (or sacred) and a negative (or demonic) version. (His particular set is specific to Judeo-Christianity and the cultures associated with it, but all story clusters seem to sprout such positive-negative pairs.) There are the Tree of Life in Paradise and its opposite, the Tree of Death that is the Cross. There are the City of God and the City of the Devil. There are the Food of Life—of which the Eucharist is the symbol—and the Water of Life, and their negative versions, the Food of Death—the fruit of the tree of knowledge in Genesis—and the Water of Death—usually in the Bible represented by the sea, which is why the Book of Revelation says, "There shall be no more sea." There is Nature as the Peaceable Kingdom or paradise where the wolf lies down with the lamb, and then there is Nature as howling wilderness, haunted by jackals presiding over ruined dwellings, which is frequently invoked in the Bible as a curse.

At the two extremes of this series of paired opposites are the heavenly realm, summing up everything we might be expected to enjoy, and the hellish one, incorporating everything evil and painful. Between these two poles stretches human life—the Merry Middle Earth of both folk ballads and *The Lord of the Rings*—and the plots of narrative literary works of all kinds show movement in one direction or another. There are narratives of fall, in which we move from the heavenly sphere to the earthly, or from the earthly to the demonic; and narratives of ascent, in which we move from hell up to earth—release-from-prison narratives are like this—or from earth up to heaven. Narratives of fall feature separation from loved ones, calamities, imprisonments, tortures, mechanical beings that mimic life, defeats, dehumanizations, and deaths; those of ascent feature reunion with loved ones, getting out of the bellies of whales, healing, nature in its more benevolent aspects, life abundant, and birth or rebirth.

As I've said, Frye was concerned mostly with literary criticism, and myths interested him as structural elements in works of literature. He used the word *myth* to mean *story*, without attaching any connotation of truth or falsehood to it; but a myth is a story of a certain kind. The myths of a culture are those stories it takes seriously—the ones that are thought to be key to its identity. Thus it was a grave matter to be told as a Canadian—as one constantly was told in the late 1950s—that one lacked a mythology. Before Leonard Cohen was a singer he was a poet, and the title of his first poetry book was *Let Us Compare Mythologies*. It was a title of its time.

Myths are not the only kinds of stories in any culture, of course. There are jokes, for example—nobody expects those to be literally true. And there are parables or fables, such as the Fox and the Grapes and the Good Samaritan. These are told to point a moral, and the moral is not dependent on the existence of a grape-loving fox or an actual Good Samaritan. "There was once" is thus a somewhat different beginning from "There was." Folk tales are stories, but few feel the need to believe in a real Aladdin or some real Twelve Dancing Princesses. There are anecdotes about old Joe down the road and Mrs. Smith across the street, or Jennifer Lopez. These are gossip. We know the people are real—they exist—but our feelings about our own inner essence—call it the soul—are not much affected by stories about them, however juicy these may be.

There are histories—stories about the past that are more or less demonstrably based on fact. (We don't know whether there was an Icarus, but we're pretty sure there was a Henry the Eighth.) These kinds of stories are viewed with considerable gravitas. People argue about histories, especially histories about their own group or country, because these really do affect how we assess our place in the world. Were we the good guys? We love people who tell us so. Were we the bad guys? We certainly hope not. If we did behave badly, why did we? Were we self-deluded, or misled by folks with ulterior motives? Did we act in good faith, or were we cynical? We need to know, because we assess our present behaviour in terms of our past behaviour, risky though such an enterprise may be, and even though we—individually—may have been born years or decades after the events in question. With history, something is at stake; it matters to us whether this or that version of "our" history is true because we think we're being told something about ourselves. Never mind that "we" were not actually present, at the First World War, or the Second, or the bombing of Hiroshima: everyone has a fully expandable ego-image, and in such matters we extend our ego beyond our own body, family, house, car, city, and state to include our country, past and present. That's another thing about myths: they gather in and circumscribe their target audience. They make a collection into a collective.

.....

Ancient myths precede histories and were once thought to *be* histories. They were thought to be true accounts of important matters. As SF often shares many features of such myths, we might want to consider these important matters. Here are some of the questions both SF and myths can pose, with some of the answers myths have provided.

Where did the world come from?

Origin myths can be divided roughly into worlds created by a sexual act of some kind (the Earth hatched from an egg, Father Sky copulated with Mother Earth), worlds that were moulded (an aquatic animal or bird dove down and brought up some mud, and the world was made from that), and worlds sung or danced or spoken into being by gods, as in *Let there be light.* One of my favourites is the Mayan creation myth. The gods created the world, but first they worried. They worried a lot. They worried and worried and worried. I'm with them.

Where did people come from?

They were fashioned from dirt, or ribs. They came from stones. They were hatched out of an egg. Mother Earth gave birth to them. The gods created them as playthings and slaves. They were baked like cookies.

Where did OUR people come from?

Dragon's teeth. Out of a clam. Descended from the Sun God. Etc.

Why do bad things happen to good people?

Because God and Satan make a bet on Job. Because the House of Atreus is cursed. Because these things are sent to try us, and God chastiseth whom He loveth. Etc.

And, more annoyingly: *Why do good things happen to bad people?*

Yes: why do the wicked prosper? Myth sometimes solves this one by an assurance that the bad people will get their comeuppance later, in the afterlife if necessary, but often things are much more ambiguous. God knows, is one answer; there must be a reason, but we see through a glass, darkly, so we don't know what the reason is.

What is right behaviour?

Lots of answers here. Under Lifestyles, everything from shoes and where they can be worn, to hairstyles, to forbidden or festive or sacred foods; and under Ethics, many choices too. As a guide to daily life, myths are slippery. We hope that good actions are rewarded and bad ones punished, and that murders and stealing and lying and cheating will be frowned upon by the story, as they are in most human societies; but in myths, it is not always so.

What do the gods want, or God, if it's a monotheism?

The answers range everywhere from your first-born to burnt kidney fat to endless devotion and obedience to not sleeping with the wrong person to having to avenge your murdered father by killing your mother. That's the problem with gods. They specialize in cleft sticks—damned if you do, damned if you don't—and they're maddeningly oblique. Gods don't come with clear instructions; or not according to the stories about them.

What are the right relationships between men and women?

Judging from the many and varied myths that tell of strenuous relationships between the sexes—huge female monster deities cut to pieces by heroic upstart-gods, women raped by immortals, mortals who are seduced by goddesses and then come to grief, gods killing guardian dragons and taking over female oracles, female demi-deities revenging themselves on faithless men, men losing paradise because women ate the apple—this seems to be uneasy ground. Stories involving gender conflict and/or separate spheres of influence—Artemis-moon by night, Apollo-sun by day—seem to be central to most mythologies. The Queen of the Night and the solar imagery in Mozart's *Magic Flute* did not come from nowhere.

. . . . .

Why were mythologies so apparently universal in what we think of as pre-literate societies? Some commentators see them as inevitable, given human grammar: if you've invented the past tense and the future tense, and if you are a question-asking being—which *Homo sapiens* is—then sooner or later the creative part of the brain is going to come up with a point of origin and an ultimate destination, even if it's the cyclic destruction and re-creation of the universe.

Early myth systems preceded writing, but once literacy spread, the old, oral mythologies were absorbed into the new medium, which at first simply recorded them—*The Iliad* existed in oral form before it was written down, we are told—and then imitated them, as Virgil did with *The Aeneid*. However, when people cease to believe that myths are literally true, believe-it-or-die theologies and perform-it-or-be-damned rituals cease to be based on them, "art" separates itself from liturgy and ritual and iconography, and myths become hidden structural principles or else the subject matter in an art that is essentially allegorical or decorative.

We may repeat the old theologies and the old rituals out of habit or a need for comfort, and we can reinterpret them in various ways; but at the same time we've been known to create new myths. Marxism and its cousin, Christian socialism, were such neo-mythical structures. Their pattern was a linear one, like that of Christianity, but for God's grand plan they substituted History, a godlike entity that would unfold in an inevitable way and justify you if you were on its right side—the side of whoever was urging you on. Then, after a number of obstacles had been overcome, a utopian society would emerge in which all inequalities and sufferings would be eliminated, much like the New Jerusalem. Here is William Morris in his inevitable-future socialist mythmaking mode:

> Come, then, let us cast off fooling, and put by ease and
> rest,
> For the CAUSE alone is worthy till the good days bring
> the best.
> Come, join in the only battle wherein no man can fail,
> Where whoso fadeth and dieth, yet his deed shall still
> prevail.
> Ah! come, cast off all fooling, for this, at least, we know:
> That the Dawn and the Day is coming, and forth the
> Banners go.
>
> —*The Day Is Coming*

Science, too, has generated new myth systems. (By "myth" I mean a story central to our self-understanding: nothing about truth or falsehood is implied.) Here, for instance, is a new creation myth: the universe began with a Big Bang. Then the Earth was formed of cosmic dust. What came before the Big Bang? A singularity. What is a singularity? We don't know.

Here is a new origin-of-people myth: people emerged via something called evolutionary forces from pre-human life forms that also so emerged. What created the rules for evolution? Life did. Where did life come from? We're not sure, but we're working on it. Why are we on Earth? No particular reason. Any idea why we should behave well? Most likely because food-gathering goes better in small groups, such as were common in the Pleistocene, if the males aren't murdering each other all the time and copulating with the other males' mates. What about relationships between men and women? We're studying that now, through brain measurements and pheromones and anthropological evidence and, oh, all sorts of things, and we hope to have a mathematical formula for falling in love any time soon. How about a belief in God, or gods? Well, granted, most cultures have had such a thing. Maybe such a belief is an evolved adaptation. Maybe your survival chances are better if you think there is a powerful being on your side who has a master plan. But there probably isn't much more to it than that.

As a story, the scientific *mythos* is not very comforting. Probably that's why it hasn't become wildly popular: we human beings prefer stories that have a central role in them for us, that preserve some of our mystery and thus some of our dignity, and that imply there might be help at hand if we really need some. The scientific version of our existence on this planet may very well be physically true, but we don't like it much. It isn't cuddly. There aren't many tunes you can hum in the shower.

. . . . .

Thus: myths are stories that are central to their cultures and that are taken seriously enough that people organize their ritual and emotional lives around them, and can even start wars over them. Such stories go underground, as it were, when the core statements about truth and reality repeated in the stories cease to be entirely, factually believed. But they then emerge in other guises, such as Art, or political ideologies.

Or films like *Avatar*. Or books like *The Left Hand of Darkness*. For every question that myths address, SF has addressed also. Indeed, it's arguable that this form and its subforms have subsumed the mythic areas abandoned by literature after the meta-theological poetics of *Paradise Lost* and the meta-theological fabulations of *The Pilgrim's Progress* and the extended theology-based other-world-building of William Blake's long "prophecies."

Before going into specifics, I'll say a little about the history of the term *science fiction*. This label brings together two terms you'd think would be mutually exclusive, since *science*—from *scientia,* meaning knowledge—is supposed to concern itself with demonstrable facts, and *fiction*—which derives from the Latin root verb *fingere,* meaning to mould, devise, or feign—denotes a thing that is invented. With *science fiction,* one term is often thought to cancel out the other. Thus such books may be judged as factual predictions, with the fiction part—the story, the characters, the invention component—rendering them useless for anyone who really wants to get a grip on, say, space travel or nanotechnology. On the other hand, they may be treated the way W. C. Fields treated golf when he spoke of it as a good walk spoiled—that is, the books are seen as fictional narratives cluttered up with too much esoteric geek material, when they should have stuck to describing the social and sexual interactions among Bob and Carol and Ted and Alice, only in futuristic clothing.

Jules Verne, a granddaddy of science fiction—in its broadest sense—on the paternal side, and the author of such works as *Twenty Thousand Leagues Under the Sea*, was horrified by the liberties taken by H. G. Wells, who, unlike Verne, did not confine himself to machines that were within the realm of possibility—such as the submarine—but created other machines—such as the time machine—that were quite obviously not. "Il invente!" Jules Verne is said to have said, with vast disapproval. He himself invented too, it must be said. But not quite so wildly.

Before the term *science fiction* became generally used, in America, in the 1930s, during the golden age of bug-eyed monsters and girls in brass brassieres, stories such as H. G. Wells's *The War of the Worlds* were labelled *scientific romances.* In both terms—*scientific romance* and *science fiction*—the science element is a qualifier. The nouns are *romance* and *fiction*, and the word *fiction* covers a lot of ground.

In the mid-twentieth century we got into the habit of calling all examples of long prose fiction "novels," and of measuring them by standards used to evaluate one particular kind of long prose fiction, namely the kind that treats of individuals embedded in a realistically described social milieu. This convention emerged with the work of Daniel Defoe—who tried to pass his inventions off as true-story journalism—and that of Samuel Richardson and Fanny Burney and Jane Austen during the eighteenth and early nineteenth centuries, and which was then developed by George Eliot and Charles Dickens and Flaubert and Tolstoy, and many more, in the mid- and late-nineteenth century.

This kind of work is found superior if it has so-called "round" characters—characters with psychological complications and moods and introspections—rather than "flat" ones who run around having narrow escapes and shooting people, round ones being thought to have more of what we call "depth." Anything that doesn't fit this mode has been shoved into an area of lesser solemnity called "genre fiction," and it is here that the spy thriller and the crime story and the adventure story and the supernatural tale and the science fiction, however excellently written, must reside, sent to their rooms—as it were—for the misdemeanour of being enjoyable in what is considered a meretricious way. They invent, and we all know they invent, at least up to a point, and they are therefore not about Real Life, which ought to lack coincidences and weirdness and action/adventure—unless the adventure story is about war, of course, where anything goes—and they are therefore not solid.

The novel proper has always laid claim to a certain kind of truth—the truth about human nature, or how people really behave with all their clothes on except in the bedroom—that is, under observable social conditions. The "genres," it is thought, have other designs on us. They want to entertain, as opposed to rubbing our noses in the daily grit produced by the daily grind. Unhappily for realistic novelists, the larger reading public likes being entertained. There's a poverty-stricken writer in George Gissing's masterpiece, *New Grub Street,* who commits suicide after the failure of his slice-of-life realistic novel entitled *Mr. Bailey, Grocer. New Grub Street* came out at the height of the craze for such adventure romance novelties as H. Rider Haggard's *She* and the scientific romances of H. G. Wells, and *Mr. Bailey, Grocer*—if it had been a real novel—would have had a thin time of it with reviewers and readers alike. If you think this can't happen now, take a look at the sales figures of Yann Martel's *Life of Pi*—pure adventure-romance—and Dan Brown's *The Da Vinci Code,* ditto, and the long-running vampiramas of Anne Rice and the *Twilight* series, and Audrey Niffenegger's *The Time Traveler's Wife.* All of them are romances rather than realistic novels proper.

The setting of the realistic novel is Middle Earth, and the middle of Middle Earth is, roughly, the middle class, and the hero and heroine are usually within the desirable norms. As publishers' readers so often say, "We *like* these people." Grotesque variations on the desirable norms appear, of course, but they take the form not of evil talking clams or werewolves or space aliens but of folks with sad character defects or strange disabilities or no incomes. Ideas about—for instance—untried forms of social organization are introduced, if at all, through conversations among the characters, or in the form of diary or reverie, rather than being dramatized, as they are in the utopia and the dystopia.

In novels proper the central characters are placed for us in social space by being given parents and relatives, however unsatisfactory or dead these may be at the outset of the story. These central characters don't just appear out of thin air as fully grown adults, the way adventure heroes are likely to do (Sherlock Holmes has no parents); rather they are provided with a past, a history. The past accounts in part for the character's inner problems, or conflicts, thus making him or her round enough to pass muster. This sort of fiction concerns itself with the conscious waking state, and if a man changes into an arthropod in such a book, he'll do so only in a nightmare.

It is up to fantasists such as Franz Kafka and Gogol to give us masterpieces in which—for instance—a man's nose becomes separated from his face and takes up independent life as a government bureaucrat, as in Gogol's "The Nose," or Gregor wakes up one morning to find he has become a beetle, as in Kafka's *The Metamorphosis*. (There is some academic literature devoted to what sort of beetle; I myself am inclined to think it was not a beetle as such but a house centipede.)

Thus, not all prose fictions are novels in the stick-to-realism sense of the word. A book can be a prose fiction without being a novel. *The Pilgrim's Progress*, although a prose narrative and a fiction, was not intended as a "novel"; when it was written, such things did not exactly exist. It's a romance—a story about the adventures of a hero—coupled with an allegory—the stages of the Christian life. (It's also one of the precursors of science fiction, although not often recognized as such.) Here are some other prose-fiction forms that are not novels proper: The confession. The symposium. The Menippean satire, or anatomy. The extended fable. And what, exactly, is *Don Quixote*? And what is *Moby-Dick*? They're stories, or they contain stories, but are they novels? In fact, the further back we stand from prose fictions—taking them all in, as it were—the fewer of them are "novels" in the nineteenth-century-realist sense of that word.

Nathaniel Hawthorne deliberately called some of his fictions "romances," to distinguish them from novels. What he might have been thinking of was the tendency of the romance form to use a somewhat more obvious degree of patterning than the novel was thought to do—the blond heroine versus her dark alter ego, for instance, as in *Ivanhoe* and the romances of Fenimore Cooper. The French have two words for the short story—*conte* and *nouvelle*—"the tale" and "the news"—and this is a useful distinction. The tale can be set anywhere, and can move into realms that are off-limits for the realistic novel—into the cellars and attics of the mind, where figures that can appear in novels only as dreams and fantasies take actual shape and walk the earth. "The news," however, is news of us; it's the daily news, as in "daily life." There can be car crashes and shipwrecks in the news, but there are not likely to be any Frankenstein monsters; not, that is, until someone in "daily life" actually manages to create one.

Fiction can of course bring us other kinds of news; it can speak, as does Yeats's golden nightingale, of "what is past, or passing, or to come." When you're writing about what is to come, you could be engaged in journalism of the dire-warning sort—elect that bastard, build that dam, drop that bomb, burn that carbon, and all hell will break loose. Such journalism is expected to confine its range to observable factors. In the nineteenth century, Tennyson wrote a poem called "Locksley Hall," which appeared to predict—among other things—the age of airplanes, and which contains the line "For I dip't into the future, far as human eye can see"; but no one can really do that. The future can never be truly predicted because there are too many variables. You can, however, dip into the present, which contains the seeds of what might become the future. As William Gibson of *Neuromancer* and cyber-punk fame said, "The future is already with us, it's just unevenly distributed." So, in forecast journalism, you can look at a lamb and make an educated guess, such as, "If nothing unexpected happens to this lamb along the way, it will most likely become (a) a sheep or (b) your dinner." But you will probably exclude (c) a giant wool-covered monster that will crush New York.

However, if your writing about the future isn't forecast journalism, it will most likely be something people will call either science fiction or speculative fiction. The terms are fluid, as we've seen. Some use *speculative fiction* as an umbrella covering science fiction and all its hyphenated forms—science-fiction fantasy and so forth—and others choose the reverse. SF novels of course can set themselves in parallel imagined realities, or long ago, and/or on planets far away. But all these locations have something in common: they don't exist, and their non-existence is of a different order than the non-existence of the realistic novel's Bobs and Carols and Teds and Alices.

Here are some of the things SF narratives can do that "novels" as usually defined cannot do.

They can explore the consequences of new and proposed technologies in graphic ways by showing them as fully operational. We've always been good at letting cats out of bags and genies out of bottles and plagues out of Pandora's Box: we just haven't been very good at putting them back in again. These stories in their darker modes are all versions of "The Sorcerer's Apprentice," in which the apprentice starts up some of the sorcerer's magic but doesn't know how to turn it off. They may help us to decide whether such apprentices could maybe use a little supervision.

They can explore the nature and limits of what it means to be human in very explicit ways, by pushing the human envelope as far as it will go in the direction of the not-quite-human. Are the robots in Čapek's *R.U.R.* human? They make a good case for their rights. Are the Stepford Wives human? How about the replicants in *Blade Runner*, or the beast folk in *The Island of Doctor Moreau*?

These are scary or creepy examples. But on the other hand, such quasi-humans can take more positive forms that help us to understand and navigate differences. In such fictions, the characters may diverge from the standard human model—Data in *Star Trek*, the gifted mutants in John Wyndham's *The Chrysalids* and the one in Russell Hoban's *Riddley Walker*, the Martians in *The Martian Chronicles*, Octavia Butler's Oankali—but they are viewed sympathetically.

SF narratives can also interrogate social organization by showing what things might be like if we rearranged them. Sometimes they are used primarily as a way of reconsidering gender structures. Charlotte Perkins Gilman's *Herland*, John Wyndham's *Consider Her Ways*, W. H. Hudson's *A Crystal Age*, the works of Joanna

Russ, Sheri Tepper's *The Gate to Women's Country,* and many of Ursula K. Le Guin's stories have this end in view.

But SF has also abounded in a subgenre we might call "economic SF"—Bellamy's industrial romance, *Looking Backward,* which anticipated the credit card, is one of these, but so is William Morris's socialistic *News from Nowhere.* Such stories, whatever else they may be doing in the way of redesigning women's clothing (sexier, less sexy) or putting food on the table (more, less; tastier, horrible), have as their central focus the production and distribution of goods and the allocation of economic benefits among various social classes.

In this respect they may use SF conventions as a semi-disguise or decorative front whereby they may criticize the present-day governments and institutions of the writer's own society when overt criticism might prove dangerous or fatal. Yevgeny Zamyatin, an early Bolshevik who saw Big Brother coming, used SF for this purpose in *We,* and Judith Merril and her generation of writer friends took to SF during the McCarthy era in the United States because they felt that bald statements of dissent would invite retribution.

Finally, SF stories can explore the outer reaches of the imagination by taking us boldly where no man has gone before, or indeed ever. Thus the spaceship, thus the inner realms of *Fantastic Voyage,* thus the cyberspace trips of William Gibson, and thus the trips between two realities in the film *The Matrix*—this last, by the way, an adventure romance with strong overtones of Christian allegory, and thus more closely related to *The Pilgrim's Progress* than to *Pride and Prejudice.*

In the process of such explorations, SF may create patterns that purport to depict the relationship of man to the universe, a depiction that takes us in the direction of religion and ultimately into the preoccupations of metaphysics and mythologies—the dispositions of gods, spirits, and demons, the origins of the universe and of the people or entities that comprise its societies, the longed-for or feared spiritual landscapes or territories, and the nature of psychic enemies. Again, this is something that can happen within the conventions of fictional realism only through conversations, reveries, stories told within stories, hallucinations, or dreams.

I'm far from the first commentator to note that science fiction is where theologically linked phenomena and reasonable facsimiles of them went after *Paradise Lost*. The form has often been used as a way of acting out a theological doctrine, as—for instance—Dante's *Divine Comedy* was once used. I'm thinking especially of C. S. Lewis's "space trilogy," *Out of the Silent Planet*, *Perelandra*, and *That Hideous Strength*, which rings the changes on the Fall of Man, Original Sin, and the possibilities of redemption, but there are by now many other examples. The religious resonances in such films as *Star Wars* are more than obvious.

Why this migration of the West's more recent founding mythologies—our once-essential core stories of the Judeo-Christian era—from Earth to Planet X? Possibly because—as a society—we no longer believe in the old religious furniture, or not enough to make it part of our waking "realistic" life. If you have a conversation with the Devil and admit to it, you're liable to end up in a psychiatric ward, not sizzling at the stake. Supernatural creatures with wings and burning bushes that speak are unlikely to be encountered in a novel about stockbrokers unless the stockbrokers have been taking mind-altering substances. But such creatures are thoroughly at home on Planet X.

So that's why Heaven and Hell—or at least some of the shapes their inhabitants have traditionally taken—have gone to

Planet X. A lot of the other gods and heroes have gone there as well. They've moved shop because they're acceptable to us there, whereas they wouldn't be here. On Planet X they can take part in a plausible story—plausible, that is, within its own otherworldly parameters. And many of us are more than willing to engage with them there because—say some theorists—our own deep inner selves still contain the archetypal patterns that produced them.

NOTES

38 Stanislaw Lem, *Memoirs Found in a Bathtub,* originally published in Cracow, 1971; Avon translation by Michael Kandel and Christine Rose. Quotation, page 10.

38 Northrop Frye, *Anatomy of Criticism*, (Princeton University Press, 1957), p. 49.

39 M. R. James: see, for instance, *Tales of an Antiquary.*

40 *Donovan's Brain* is by Curt Siodmak. *The Kraken Wakes* is by John Wyndham.

41 Jane Austen: Thus the frisson produced by such titles as *Pride and Prejudice and Zombies.*

43 Denis Dutton, *The Art Instinct: Beauty, Pleasure, and Human Evolution* (Oxford University Press, 2009).

46 Season-linked story cycle: For a good late-Victorian example, see William Morris, *The Earthly Paradise.*

56 Bob and Carol and Ted and Alice, from the film of that name.

60 Yeats's nightingale: in "Sailing to Byzantium."

63 Judith Merril: Told to this author.

Dire Cartographies: The Roads to Ustopia

What we call places are stable locations with unstable converging forces.

REBECCA SOLNIT, *Infinite City: A San Francisco Atlas*

. . . after WWII utopia was no longer just a synonym for naiveté. It was dangerous. Now, decades further on, in a new century and a new millennium, earnest utopian thought and earnest utopians are a glowing ember at best, and utopia's legion failures seem to suggest that the best course of action would be to crush it—snuff it for good.

J. C. HALLMAN, *In Eutopia*

This chapter is about literary utopias and dystopias, and how it came to pass that I found myself writing about them, and then—many years later—attempting something in that form myself. *Ustopia* is a word I made up by combining utopia and dystopia—the imagined perfect society and its opposite—because, in my view, each contains a latent version of the other.

The "dire" might at first glance appear to be connected only to the obverse or dystopic side of this coin, where unpleasantness prevails, though most utopias viewed slantwise—from the point of view of people who don't fit into their high standards of perfection—are equally dire. But before expanding on that notion, I'll begin with the second word of my title: *cartographies.*

Cartography is map-drawing, and the brain is, among other things, a map-making entity. Not only our brains, but also those of other animals with brains. And not only the brains of other animals with brains: even the lowly slime mould, entirely devoid of a central nervous system, "maps" its adjacent space, figuring out—for instance—the closest route to enjoyable foods. (It prefers oatmeal.) From our earliest days, as soon as we can crawl around on the floor, we are inscribing maps of our surroundings onto the neural pathways in our brains and—reciprocally—inscribing our own tracks, markings, and namings and claimings onto the landscape itself. Snails make trails, and so do beavers, and so do tree-scratching bears and hydrant-marking dogs, and so—quintessentially—do we. We're almost as good at finding the shortest path to enjoyable foods as slime moulds are, although our choice may not be oatmeal.

With every map there's an edge—a border between the known and the unknown. In old medieval and early Renaissance maps, the edges were where the monsters were drawn—the sea serpents and many-headed hydras, which were, as we say, off the map. Monsters live under the bed when you're little because you can't see under the bed when you're actually *in* the bed. And that's what's scary about darkness for a lot of people: the unknown. The known is finite, the unknown is infinite: anything at all may lurk in it. Grendel, the monster in *Beowulf,* is termed—in John Gardner's reworking of him—"earth rim roamer" and "walker of the world's weird wall." That's where monsters live—at the edges, at the borders. Monsters also live at the edges of our consciousness—during the day and in stable times, that is. They take full possession of our field of vision only when we're "asleep" or entranced in some way, as suggested by the title of Goya's enigmatic engraving of 1799, *The Sleep of Reason Breeds Monsters.*

Why do we so frequently put monsters at the edges of the maps, or under the bed, and also—in some of its forms—in stories of the adventure-romance type? As Roberto Calasso sagely notes in *The Marriage of Cadmus and Harmony,* heroes need monsters in order to establish their heroic credentials through combat, but monsters most emphatically do not need heroes. And once the monsters are slain, the heroes die as well, to be replaced by urban planning with its crisp maps of the known, and then, in the shadow world hidden or possibly pushed aside by those maps, by more monsters, who are endlessly re-created so that yet more heroes—detectives, spies, police officers, secret agents, and their like—may do battle with them.

The edges of the maps—out of our sight, beyond the known world—were also where the writers of early utopias set their tales. During the Middle Ages, utopias were not devised much, since the perfect society had been postponed to the life after death or until after the millennium or the second coming; as the 1930s Wobbly folk song put it, "There'll be pie in the sky when you die." There was a no-place called The Land of Cockaigne, where the walls were made of pies and the shingles of cakes, sex was unrestricted, and laziness and gluttony were available to all; but although a paradise of sorts, it was—officially at any rate—a fools' paradise.

However, once the Renaissance and then the early modern age got going, utopia made a comeback. Like Plato's seminal Atlantis and the Avalon of Arthurian romances, these utopias were typically located on islands to be found just out of reach of the real maps, like the utopia in the book of that name by Thomas More.

Even the deserted island of Shakespeare's *The Tempest* contains a utopia: the Golden Age society described by the kindly character Gonzalo, where no one has to work, where all are free and equal by decree of the king—a slight contradiction there—and where there is no crime or war. (The dystopia latent in *The Tempest* is the very same physical location viewed through the eyes of Caliban, its original inhabitant.) In Jonathan Swift's *Gulliver's Travels*, the utopias are comic and satirical, but they are similarly located on islands, each of which is provided with a realistic-looking map by the fictional wandering and tale-telling sailor, Lemuel Gulliver—in the tradition of an earlier sea captain and an earlier map, those in More's *Utopia*.

But then real mapping filled in previously "undiscovered" areas of the ocean, thus ruling out islands and driving utopia-dystopia further into the unknown. First it went underground, to the traditional location of under-the-hill fairylands and worlds of the dead and the kinds of mountain-king dwarfs we find in Tolkien and in the folklore he drew on. (The underworld is also the unknown realm of choice for various concocted nineteenth-century gnome-lands and fairylands, that of Lewis Carroll's *Alice* among them.) Such cavern sagas as Bulwer-Lytton's *The Coming Age* and Jules Verne's *Journey to the Center of the Earth* were placed in vast hollows beneath the Earth's crust, replete with surviving prehistoric beasts and giant ferns.

But then, once the Earth's structure had been more fully described by geologists, ustopia moved to unexplored hinterlands, where we find H. Rider Haggard's lost city of Kôr in his novel *She*, or the Shangri-La of James Hilton's *Lost Horizon*, or H. G. Wells's "Country of the Blind." But these locations also became too thoroughly mapped, and ustopia had to relocate again.

For a while the other planets in the solar system were possibilities, though Mars and Venus and the Moon had to be given up once we knew what was actually on them, such as no intelligent life. The final move was to an outer space far beyond our system, or to a parallel universe, or to a past so long ago that all traces of it have been obliterated; or to the future, also an unknown.

In the previous chapter, "Burning Bushes," I suggested that the literary offspring of theology, such as angels and devils, moved to outer space because we no longer believed in their doctrinal underpinnings sufficiently to make these creatures plausible in realistic narratives set on Earth. But maybe this emigration was also caused by a real estate problem. We filled the unknown spaces with us—with ourselves, and our names and roads and maps. We tidied up, we gentrified, we put in streetlights; so the rowdy and uncontrollable bohemians of the imagination—always dwellers in the penumbras—had to move on.

Maps are not only about space, they're also about time: maps are frozen journeys. They may be journeys from the past: places we've been, or whose history we're studying. (Who can begin to comprehend the Second World War without those arrow-covered maps?) Or they can be journeys of the present, helping you to find your way to the nearest organic coffee shop via the "route" button on your GPS-enabled phone. Or they may be journeys of the future, by the aid of which you plan your next vacation—how you'll get to the island of your dreams, how long that will take, where you'll go while you're there, how you'll get back.

The many fictional inner journeys available to us—those that unfold in imaginary places—also come equipped with maps. Think of the mandatory maps at the fronts of those 1930s country-house murder mysteries, with the library and the conservatory and the servants' wing, or the maps included in Ursula K. Le Guin's *Earthsea Trilogy* and those in *The Lord of the Rings*; and more, and more. Indeed, a great many more. It does seem to be a rule as well that when there are wars or murders involved—anything requiring strategic planning and the movement of opponents or aggressors toward their targets—maps are a conceptual help, not only for the reader but for the writer.

Indeed it would seem that quite a few writers think cartographically, especially writers about imaginary places. If you're writing about a real city, a well-known one, the maps of it already exist and the reader can look them up, but if you're writing about an unknown location, they don't. The writing of *Treasure Island* began with Stevenson's drawing of a map, as an effort to amuse a young visitor. It was not until after he'd sketched in some buried treasure and a few landmarks that Stevenson began writing the story proper, and the book's journey began with that very same map being discovered in the sea chest of dead pirate Billy Bones.

As ustopia is by definition elsewhere, it is almost always bracketed by two journeys: the one that transports the tale-teller to the other place and the one that transports him (or her) back so he can deliver his report to us. Thus the writer of the book always has to come up with a mode of transport. When utopias were placed on islands, the journey was a simple matter of a sea voyage, and then of some sort of rescue by boat. Journeys underground involved tunnels, and ropes, and falling down holes, and the sudden breaking through of stone walls; return journeys took luck, scrambling up cave walls, the following of an animal that knew an escape route, or a version of Ariadne's thread. Locations in outer space necessitated spaceships.

As for journeys to the future, which require transport not

through space but through time, one could always fall back on that medieval gimmick, the dream vision, a form of psychic teleportation; or some kind of time machine; or a long sleep, like Sleeping Beauty's or Rip Van Winkle's. (*Looking Backward* and *A Crystal Age* both use this one: in the latter, our time-travel reporter bangs his head and wakes up far in the future, charmingly covered with little tree roots.) It's this convention that Woody Allen is satirizing in his film *Sleeper* when he crawls out of a freezer covered in tinfoil.

Once "the future" became an established location, writers could feel free to jettison the travel episode and the "reporter" figure, and to plop the reader right down in the midst of things. "It was a bright cold day in April," begins *Nineteen Eighty-Four,* "and the clocks were striking thirteen." The "reporter" function exists in the book not as a person but in two texts within it: a book forbidden by the ruling Party of 1984, *The Theory and Practice of Oligarchical Collectivism,* by Party arch-enemy Emmanuel Goldstein—who may or may not exist—and "The Principles of Newspeak," an essay on language-as-control that we read after the end of the story proper. It is this essay by an unknown reporter—in my belief—that travels to our own time and lets us know how things turned out.

In Aldous Huxley's *Brave New World,* the "reporter" is replaced by the "Savage," a man from outside the borders of the highly organized tech utopia whose views on life may be thought to have more in common with those of the reader—or at least the reader of the 1930s, when *Brave New World* was written. The Savage plays, in effect, a tragic Caliban to all the cheery Mirandas and hidden Prosperos in the book—the perfumed anti-pregnancy free-sex girls, the largely unseen string-pullers that direct and regiment the society in order to protect people from themselves.

Reporters and messages naturally require some means of transmission. Perhaps no other genre has so concerned itself with information systems as has SF, especially that of the ustopian kind. Various methods have been employed by various authors. There are diaries and journals left by the literary descendants of Robinson Crusoe in the hope that someone in the future may read them; there are strange manuscripts found in copper cylinders; there are metal books, and crystal encoding systems, and hieroglyphs that need deciphering. There are language barriers to be overcome, there are catastrophes resulting in mass memory and information loss—Stanislaw Lem's *Memoirs Found in a Bathtub* postulates a worldwide meltdown caused by a nanobioform that eats paper, for example, thus creating the equivalent of a worldwide library-burning.

Some writers just skip the message-transmission system and resort to straight third-person narration, or allow the narrator to address us from limbo. But any writer of ustopias has to answer three necessary questions: where is it, when is it, and—in relation to maps—what shape is it? For unless we readers can believe in the ustopia as a potentially mappable place, we will not suspend our disbelief willingly.

.....

I came early to maps, though not altogether by choice. My older brother was an inveterate map-maker. Not only did he devise follow-the-clue maps for me, but he drew many maps of imaginary places on other planets. The lands he described were often islands, as is the real estate for sale on the virtual-reality site Second Life—islands are more comprehensible and easily defined than countries with contingent borders. While mapping Neptune and Venus, my brother also took to mapping the very island we were then living on, naming each bay, swamp, promontory, peninsula, and offshore island. Once those places had names, it was, strangely enough, easier to find your way to them.

Naming is of course an aid to memory: attach a name to a place and you have a proto-map. The physical maps are only the outward and visible manifestations of inner and neurological maps—the things drawn or, in the case of the northern Canadian Inuit, carved, in three dimensions on pieces of wood that could float if your kayak tipped. And, as with anything brainiac, practise makes bigger: those who study the brains of taxi drivers in London, where apprentice taxi drivers have to learn the city by memory and then pass a difficult test, report that the mapping areas of the taxi drivers' brains—those parts having to do with spatial orientation and visualization—are larger and denser than those in the brains of the rest of us.

In addition to being, almost always, a mapped location, ustopia is also a state of mind, as is every place in literature of whatever kind. As Mephistopheles in Marlowe's play *Doctor Faustus* tells us, Hell is not only a physical space. "Why this is Hell, nor am I out of it," he says:

Hell hath no limits, nor is circumscrib'd
In one self place; but where we are is hell,
And where hell is, there must we ever be.

Or, to cite a more positive version, from Milton's *Paradise Lost*:

. . . then wilt thou not be loath
To leave this Paradise, but shalt possess
A Paradise within thee, happier farr.

In literature, every landscape is a state of mind, but every state of mind can also be portrayed by a landscape. And so it is with ustopia.

. . . . .

How did I myself come to create my own ustopias—these not-exactly places, which are anywhere but nowhere, and which are both mappable locations and states of mind?

It was an indirect journey. I'd decided that I was a writer when I was sixteen, after spontaneously generating a pretty bad poem that I thought was pretty good. I was in the twelfth grade at the time, without a living-writer role model in sight. I did not have the foggiest notion of how to go about being a writer, though it was clear to me that I would need a day job, at least at first, since even such an optimistic person as myself could not expect to burst into best-sellerdom at once.

Our generation did do some writing in school, but it was in the form of essays, or else grammar and composition. We were not encouraged to write fiction and poetry, although we did read a lot of these. Should we be overtaken by the Muse, we could always publish the results in the school yearbook, if we had no shame.

After taking a couple of false turns—luckily, only in my

head—I elected to go to university after all. (I had, briefly and madly, decided I would support myself by writing True Romance stories. This seemed easy enough, as they were all basically some variation of *Wuthering Heights,* in which the girl wrongly falls for the guy with the motorcycle instead of the one with the steady job at the shoe store. But I found I couldn't do this: as with any kind of writing, you somehow have to believe in it yourself or it isn't convincing.)

Then I had a short period of thinking I might become a journalist. But a second cousin who was in fact a journalist—he'd been dredged up by my parents in order to discourage me about the newspaper life and herd me in the direction of higher education—told me that women journalists only ever wrote the ladies' pages and the obituaries, and my—by then—snobby and bohemian self cringed in horror.

Off to university I duly went; but after four years of Honours English, the question of what to do next once more became a pressing one. By this time I was if anything even more bohemian, and was already a coffee-house reader of my still rather terrible poetry, so I thought I should go to London, or possibly to Paris, and live in a cockroach-infested garret, and write masterpieces while gnawing crusts of bread and, if I was really up to it, drinking absinthe. But I was again headed off by my benevolent elders: I was urged to apply for a scholarship to Harvard, where—I was assured—I would probably be able to get more writing done than I would while shivering in the garret and would anyway come out with a job ticket, and therefore be able to write my deathless masterpieces during the long, leisurely summer vacations known to be enjoyed by college professors.

So I decided to postpone the absinthe-drinking, and I did get the scholarship, and I soon found myself in the land of my ancestors, which—in part—is Puritan New England. There, beginning in 1961, I studied Victorian Literature—ask me anything about the Freudian implications of Edward Lear's poem "The Pobble That Has No Toes" and you will receive a long and pedantic answer. This was a time when Victorian literature was just beginning to recover from the disdain heaped upon it by the modernist likes of Lytton Strachey and T. S. Eliot; when the Pre-Raphaelite paintings that were the vogue in the late 1800s were stacked in the back rooms of the Fogg Museum, having not yet achieved their present-day apotheosis as picture postcards; and when Oscar Wilde's pronouncement on Dickens—"One must have a heart of stone to read the death of Little Nell without laughing"—was the general literary opinion. The serious, indeed the orthodox thing to study was the metaphysical poets, like John Donne, or the contemporaries of Shakespeare, like Webster and Marlowe. But I have always had a less than orthodox side.

In addition to the Victorians, I took courses in American Literature and Civilization because I was told it was my "gap"—one I needed to fill in order to write the required comprehensive exams. We hadn't heard much about Cotton Mather or John Winthrop or "The Day of Doom" by Michael Wigglesworth up in Canada, worse luck. But that gap was soon filled: ask me anything about the Salem Witch Trials and the rules of spectral evidence, and you will receive an even longer and more pedantic answer.

Being a confirmed rummager, I enjoyed all of this meandering around in the sidebars of literary history, even though I was not allowed into the Lamont Library, where all the modern poetry was kept, on account of being a girl; but I compensated in the stacks of the Widener, which had everything you might want to know about demonology. In those stacks there were more obscure books than you could ever hope to find elsewhere, even on the Internet today, and I whiled away many a misspent hour reading about things that were none of my business—the Widener stacks being a much bigger version of the book-filled cellar of my parents' house where I used to avoid doing my homework.

Having duly passed my Orals, I had to decide on a thesis topic. Dreaded quest! Your thesis was supposed to be about something that hadn't yet been, as they say, *done,* and when it came to the major writers, such topics seemed few and far between.

It was now that my earlier reading in non-canonical literature came to my aid. At first I thought I would write about W. H. Hudson, whose lyrical novel, *Green Mansions,* seemed worthy of investigation. In it there is an otherworldly girl called Rima who belongs to an anthropological group of one, and can talk to birds and animals, and gets burnt up in a giant Tree-of-Life by hostile Indians. But I soon expanded my scope to include a line of literary descent that led from the earlier Scottish writer George MacDonald—author of, among other things, *At the Back of the North Wind,* which had captivated me as a child—through H. Rider Haggard's highly influential book *She,* all the way to the non-realistic prose fictions of C. S. Lewis and J. R. R. Tolkien. I should point out that at that time no one of any academic respectability was paying any attention to this kind of writing, or to "science fiction" and its related forms or subforms, such as fantasy and ustopias. Lewis and Tolkien had come out of academia, but they had not yet been accepted back into it as writers, so I was on my own. However, Marshall McLuhan's famous dictum "Art is what you can get away with" hadn't been lost on me, and I saw no reason why it shouldn't apply to Ph.D. theses as well.

I called my thesis "The English Metaphysical Romance" because the books I was studying included other-than-human beings and treated themes that were, in origin and in subtext, theological in nature. Someone once said that such works could only have been produced by Anglicans, no longer Catholic but not exactly Protestant either: the metaphysical romance was where the "real presence" went—the magic, transformational part of the Eucharist changed bread and wine into flesh and blood—once the Anglicans had renounced its factuality and turned it into a symbol.

Others considered this type of fantasy writing the result of repression in an age that censored any overt mention of sexuality. One related by-product of this repression was the unhealthy Victorian obsession with fairy paintings, showing Titania and her train, revels near giant mushrooms, and related scenes—basically a method of slipping past Mrs. Grundy in order to paint naked people having orgies. Orgies were apparently acceptable if you made the naked or semi-clad people very small and put butterfly wings on them. "I hate fairies," one of my English friends said to me recently. "Nasty little pink wriggly things!" It's true, many of the painted Victorian fairies *were* little and wriggly, though some of them were blue rather than pink. Others, however, were more goddesslike; one could see the connection between Fairy Queens—long lustrous hair, diaphanous draperies and all, and hardly ever a Fairy King in sight—and the kinds of larger-than-life Queen Bee female figures that were rapidly taking centre stage in my thesis.

These powerful female figures in the "metaphysical romances" I was studying were not goddesses, but they were not normal human women either. What then were they, besides being the great-grannies of Wonder Woman? To this question I now consecrated my waking hours. I divided the thesis into two parts: "The Power of Nature," in which I explored two kinds of strong supernatural female figures—good ones, which I saw as Wordsworthian nature deities of the "Nature-never-did-betray-the-heart-that-loved-her" variety, and bad or morally ambiguous ones, which I saw as Darwinian, or the red in tooth and claw alternate species. George MacDonald's North Wind and young-old grandmother figures—to his mind Christian allegories of Grace and the like—I saw as exemplifying the "good" kind, while H. Rider Haggard's *She* represented the Darwinian kind—not evil as such but amoral.

The second part of the thesis was called "The Nature of Power" and was devoted to an examination of the different kinds of societies associated with these two types of female figures—"good" societies, which were always connected to jolly agriculturalists like the hobbits and/or with woodland activities like those of the elvish folk headed up by Galadriel in *The Lord of the Rings*, and "bad" societies, which were not only disagreeable tyrannies full of Orcs and other nasties but highly industrialized and polluting. The bad societies were destructive toward nature and its creatures, especially trees, thus giving us one of the most satisfying scenes in *The Lord of the Rings*: the revenge of the treeish Ents. (Though in Tolkien's work, as in many fictional worlds from *The Wonderful Wizard of Oz* to Harry Potter, there are hostile trees as well.)

Thus when I went to see the movie *Avatar* I knew exactly where I was. I was (a) in a Royal Academy exhibition called Victorian Fairy Painting, with the giant luminous plants, the scantily clothed people with big ears, and so forth; and (b) in my own thesis of the 1960s, with the sin of Tree-of-Life burning, the supernatural female figures, the bad machine-makers and forest-despoilers, and the whole ball of wax.

I never finished my thesis, since I got diverted by novel-publication and film-script writing around 1969–70; but in the course of riffling through obscure books that, at that time, nobody but me was interested in, I discovered lots and lots of utopias. The nineteenth century, especially the second half of it, was so cluttered up with them that Gilbert and Sullivan wrote a parody operetta called *Utopia Limited.* I also discovered—beginning around the turn of the century but gathering steam as the twentieth century progressed, if *progressed* is the word—a strain of increasingly darker and more horrifying dystopias.

Why this change? In the nineteenth century, there had been many rapid technological, scientific, and medical changes—improved sewer systems and sanitation, antiseptics such as carbolic acid, anesthesia, vaccination, advances in transportation and manufacturing, and many more. The future looked set to continue in this ever-rosier direction, or as Tennyson's fiery young idealist put it in his poem "Locksley Hall," "Let the great world spin forever down the ringing grooves of change." (The metaphor came from trains, but Tennyson hadn't looked very closely at the tracks: he thought they were concave.)

The nineteenth century's positive utopias were inspired, as well, by various radical social thinkers, including William Cobbett and Karl Marx, and by Christian socialists such as Charles Kingsley and John Ruskin. Many people still really believed that humankind was almost perfectible if only society could change the way it was organized. People wrote utopias—such as William Morris's arts-and-crafts socialistic *News from Nowhere* or Edward

Bellamy's technically advanced utopia *Looking Backward*—because they really did think that humankind could do better than the inequality, social injustice, vice, dirt, disease, and squalor the writers witnessed all around them. Their utopias are versions of the Before-and-After makeovers you used to see in women's magazines. Before, a sloppy, sad, run-down failure; but add a nifty haircut, flattering wardrobe, more healthful diet, well-applied eyeshadow, and look! A smiling, energetic, and sexier whole different person! (Though if the whole different person is smiling too eerily, watch out: you may be in a dystopia after all because, like the how-may-I-help-you women in *The Stepford Wives,* that person may be a robot.)

Along with the literary utopias, the nineteenth century spawned hundreds of actual ones—groups of people who set up new communities—from the socialist Finnish colonies on the west coast of Canada, to the Esperanto-speakers who thought that a universal language might result in world peace, to the Oneida Community that practised an intricate form of polygamy and morphed into a flatware company. These had as their ancestors a large number of utopian religious communities, ranging from the Quakers, a disruptive cult that would sometimes streak church gatherings before it settled down into the more sober-sided version that went in for oatmeal and prison reform, to the Shakers (who did away with sex—oddly enough, they have died out), to the Mennonites and Amish.

The seventeenth-century Puritan New Englanders began, too, as utopianists. The phrase "a city upon a hill, a light to all nations" may sound familiar, since it was used recently by an American president, but it was first attached to America in the seventeenth century by John Winthrop, and comes from the inspiring utopian prophecy in the Book of Isaiah crossed with a sermon by Jesus. The New England Colony saw itself as the City of God in action—like so many utopias, it was going to start again and do things right this time. However, as Hawthorne pointed out, the first public-works items the colony built were a prison and a scaffold—acknowledgements of its own dystopic underbelly.

The nineteenth-century literary utopias concentrated less on religious structures and more on material improvements, but the lustre of both physical and spiritual utopian light dimmed considerably in the twentieth century. Despite this dimming, in the burst of Edwardian splendour that preceded the First World War there were brilliant upflarings of utopianism in the world of art, now dubbed, collectively, "utopian modernism." These European art movements wished not merely to reflect the world but to change it. Under this heading we find everything from Italian Futurism, the Bauhaus, De Stijl, and Russian Constructivism: all wanted to overthrow established ideas and conventions and set up their own new and improved versions.

Though utopian from their own point of view, some of these movements are dystopian from ours; indeed, in their frequent celebration of violence, they point to a recurring motif in literary as well as in political utopian thinking: the brave new order often comes about as the result of war and chaos.

Then along came the real war—the Great War—which did change the world but at horrifying cost. And then, in this changed but not improved postwar world, several societies had a chance to practise utopian social engineering on a large scale. Most noteworthy were the U.S.S.R. under Lenin and Stalin and Germany under Hitler. The result, in each case, was unprecedented bloodshed and the ultimate collapse of the supposedly utopian system.

Lest we assume that communists and fascists were the only sorts of thinkers to go in for this sort of thing, there are many lesser-known entries in the list of failed utopias, including a capitalist-and-workers' paradise set up by Henry Ford in the 1920s and 1930s. It was called "Fordlandia," after its founder, and has recently been the subject of two books, both of them called *Fordlandia*: a factual account by Greg Grandin and a novel by Eduardo Sguiglia. Fordlandia was situated in the backwoods of Brazil, where the happy workers were supposed to grow rubber trees to make tires for Henry Ford's Fords; but despite urban planning, and swimming pools for management, and despite or perhaps because of Ford's efforts to regiment all employees and turn them into teetotallers like himself, the community soon fell apart in a welter of corruption, waste, vice, snakebites, tropical diseases, violence, and rebellion.

Why is it that when we grab for heaven—socialist or capitalist or even religious—we so often produce hell? I'm not sure, but so it is. Maybe it's the lumpiness of human beings. What do you do with people who somehow just don't or won't fit into your grand scheme? All too often you stretch them on a Procrustean bed or dig a hole in the ground and shovel them into it. With so much stretching, hole-digging, and shovelling going on as the twentieth century ground on, it was difficult to place faith in the construction of utopias, literary or otherwise. It became much easier to depict awful societies not as the tawdry Before side of an After happy-face future but as the much worse thing we might instead be heading toward. The future societies imagined by mid- and late-twentieth-century writers, and indeed by early-twenty-first-century ones, are much more likely to be dark than bright.

. . . . .

Throughout this chapter I've been using the term *ustopia,* and now I'll expand on it. Utopia, as you know, comes from Thomas More's book of that name—which in his case may mean either "no place" or "good place," or both. Some are of the opinion that More's book was a sort of joke: utopia can't exist because fallen human nature doesn't permit it. Nevertheless, his term stuck, and now, by general usage, utopias are thought to portray ideal societies or some version of them. Their program is to do away with the ills that plague us, such as wars, social inequality, poverty and famine, gender inequalities, fallen arches, and the like. (People—especially women—are always better looking in nineteenth-century utopias than the authors thought they were in real life.)

Dystopias are usually described as the opposite of utopias—they are Great Bad Places rather than Great Good Places and are characterized by suffering, tyranny, and oppression of all kinds. Some books contain both—a sort of "look on this picture, then on that," as Hamlet puts it—one, noble and virtuous; the other, corrupt and vicious. Polar opposites.

But scratch the surface a little, and—or so I think—you see something more like a yin and yang pattern; within each utopia, a concealed dystopia; within each dystopia, a hidden utopia, if only in the form of the world as it existed before the bad guys took over. Even in Orwell's *Nineteen Eighty-Four*—surely one of the most unrelievedly gloomy dystopias ever concocted—utopia is present, though minimally, in the form of an antique glass paperweight and a little woodland glade beside a stream. As for the utopias, from Thomas More onwards, there is always provision made for the renegades, those who don't or won't follow the rules: prison, enslavement, exile, exclusion, or execution.

.....

Forty years after having abandoned my "metaphysical romance" thesis with its chapters on good and bad societies, I find I have produced—so far—three novel-length ustopias of my own: *The Handmaid's Tale*, *Oryx and Crake*, and *The Year of the Flood*.

Why did I jump the tracks, as it were, from realistic novels to dystopias? Was I slumming, as some "literary" writers are accused of doing when they write science fiction or detective stories? The human heart is inscrutable, but let me try to remember what I thought I was up to at the time.

First, *The Handmaid's Tale*. What put it into my head to write such a book? I had never done anything like it before: my previous fiction had been realistic. Tackling a ustopia was a risk. But it was also a challenge and a temptation because if you've studied a form and read extensively in it, you often have a secret hankering to try it yourself.

I began the book—after a few earlier dry runs—in Berlin in the spring of 1984. I had a D.A.A.D. fellowship, in a program run by West Berlin to encourage foreign artists to visit, as the city was at that time encircled by the Berlin Wall and its inhabitants felt understandably claustrophobic. During our stay we also visited East Berlin, as well as Poland and Czechoslovakia, and I thus had several first-hand experiences of the flavour of life in a totalitarian—but supposedly utopian—regime. I wrote more of the book once I was back in Toronto, and completed it in Tuscaloosa, Alabama, in the spring of 1985, where I was the holder of an M.F.A. chair. Tuscaloosa and Alabama provided another kind of flavour—that of a democracy, but one with quite a few constraining social customs and attitudes. ("Don't ride a bicycle," I was told. "They'll think you're a communist and run you off the road.")

The writing of *The Handmaid's Tale* gave me a strange feeling, like sliding on river ice—exhilarating but unbalancing. How thin is this ice? How far can I go? How much trouble am I in? What's down there if I fall? These were writerly questions, having to do with structure and execution, and that biggest question of all, the one every writer asks him- or herself with every completed chapter: *Is anyone going to believe this?* (I don't mean literal belief: fictions admit that they are invented, right on the cover. I mean, "find the story compelling and plausible enough to go along for the ride.")

These writerly questions were reflections of other, more general questions. How thin is the ice on which supposedly "liberated" modern Western women stand? How far can they go? How much trouble are they in? What's down there if they fall?

And further: If you were attempting a totalitarian takeover of the United States, how would you do it? What form would such a government assume, and what flag would it fly? How much social instability would it take before people would renounce their hard-won civil liberties in a tradeoff for "safety"? And, since most totalitarianisms we know about have attempted to control reproduction in one way or another—limiting births, demanding births, specifying who can marry whom and who owns the kids—how would that motif play out for women?

And what about the outfits? Ustopias are always interested in clothing—either less of it, compared to what we wear now—that was popular in Victorian times—or more of it, compared to what we wear now. The clothing concerns usually centre around women: societies are always uncovering parts of women's bodies and then covering them up again. (Maybe this is just to keep things interesting: now you see it, now you don't, though the "it" changes a lot. What was it that used to be so very alluring about a trim set of ankles?)

My rules for *The Handmaid's Tale* were simple: I would not put into this book anything that humankind had not already done, somewhere, sometime, or for which it did not already have the tools. Even the group hangings had precedents: there were group hangings in earlier England, and there are still group stonings in some countries. Looking further back, the Maenads, during their Dionysian celebrations, were said to go into frenzies during which they dismembered people with their hands. (If everyone participates, no one individual is responsible.) For a literary precedent, one need search no further than Emile Zola's *Germinal,* which contains an episode in which the town's coal-mining women, who have been sexually exploited by the shopkeeper, tear the man apart and parade his genitalia through town on a pole. A less raw but still shocking precedent is Shirley Jackson's short story "The Lottery" (which I read as a teenager, shortly after it came out, and which made a chilling impression on me).

The coverups worn by the women in *The Handmaid's Tale* have been variously interpreted as Catholic (as in nuns) or Muslim (as in burkas). The truth is that these outfits are not aimed at any one religion. Their actual design was inspired by the Old Dutch Cleanser figure on the sink cleaner boxes of my childhood, but they are also simply old. Mid-Victorians, with their concealing bonnets and veils to keep strange men from leering at their faces, would not have found them so unusual.

I prefaced the novel with three quotations. The first is from the Bible—Genesis 30, 1:3, the passage in which the two wives of Jacob use their female slaves as baby-producers for themselves; this ought to warn the reader against the dangers inherent in applying every word in that extremely varied document literally. The second is from Jonathan Swift's *A Modest Proposal*: it alerts us to the fact that a straight-faced but satirical account—such as Swift's suggestion that the grinding Irish poverty of his times could be alleviated by selling and eating Irish babies—is not a recipe. The third—"In the desert there is no sign that says, 'Thou shalt not eat stones' "—is a Sufi proverb stating a simple human truth: we don't prohibit things that nobody would ever want to do anyway, since all prohibitions are founded upon a denial of our desires.

The Handmaid's Tale was published in Canada in the fall of 1985, and in the United States and the United Kingdom in the spring of 1986. In the United Kingdom, its first reviewers treated it as a yarn rather than a warning: they'd already experienced Oliver Cromwell and his Puritan republic and seemed to have no fear of re-enacting that scenario. In Canada, people asked, in anxious Canadian fashion, "Could it happen here?" In the United States, Mary McCarthy, writing in the *New York Times*, gave the book a largely negative review on the grounds that it lacked imagination, and anyway it was unlikely ever to take place, not in the secular society she perceived as the American reality. But on the West Coast, so attuned to earthquake tremors, switchboards on talk shows lit up like Las Vegas, and someone graffitied on the Venice Beach seawall: "*The Handmaid's Tale* Is Already Here!"

It wasn't already here, not quite, not then. I thought for a while in the 1990s that maybe it never would be. But now I'm wondering again. Of recent years, American society has moved much closer to the conditions necessary for a takeover of its own power structures by an anti-democratic and repressive government. Approximately five years after *The Handmaid's Tale* was published, the Soviet Union disintegrated, the West slapped itself on the back and went shopping, and pundits proclaimed the end of history. It looked as if, in the race between *Nineteen Eighty-Four* and *Brave New World*—control by terror versus control through conditioning and consumption—the latter had won, and the world of *The Handmaid's Tale* appeared to recede. But now we see a United States weakened by two draining wars and a financial meltdown, and America appears to be losing faith in the basic premises of liberal democracy. After 9/11, the Patriot Act passed with barely a cough, and in Britain citizens have accepted a degree of state supervision that would once have been unthinkable.

It's a truism that enemy states tend to mirror one another in organization and methods. When colonies were the coming thing, everyone wanted one. Atom bombs in the United States created the desire for some in the U.S.S.R. The Soviet Union was a large bureaucratic centralized state, and so was the America of those times. What form will the United States assume now that it's opposed by unrelenting religious fanaticisms? Will it soon produce rule by the same kind of religious fanaticism, only of a different sect? Will the more repressive elements within it triumph, returning it to its origins as a Puritan theocracy and giving us *The Handmaid's Tale* in everything but the outfits?

I've said earlier that dystopia contains within itself a little utopia, and vice versa. What, then, is the little utopia concealed in the dystopic *Handmaid's Tale*? There are two: one is in the past—the past that is our own present. The second is placed in a future beyond the main story by the Afterword at the end of the book, which describes a future in which Gilead—the tyrannical republic of *The Handmaid's Tale*—has ended, and has thus become a subject for conferences and academic papers. I suppose that's what happens to ustopian societies when they die: they don't go to Heaven, they become thesis topics.

.....

After *The Handmaid's Tale* there was a period of approximately eighteen years during which I did not write ustopian novels, but then came *Oryx and Crake* in 2003. *Oryx and Crake* is dystopic in that almost the entire human race is annihilated, before which it has split into two parts: a technocracy and an anarchy. And, true to form, there is a little attempt at utopia in it as well: a group of quasi-humans who have been genetically engineered so that they will never suffer from the ills that plague *Homo sapiens sapiens.* They are designer people. But anyone who engages in such design—as we are now doing—has to ask, How far can humans go in the alteration department before those altered cease to be human? Which of our features are at the core of our being? What a piece of work is man, and now that we ourselves can be the workmen, what pieces of this work shall we chop off?

The designer people have some accessories I wouldn't mind having myself: built-in insect repellant, automatic sunblock, and the ability to digest leaves, like rabbits. They also have several traits that would indeed be improvements of a sort, though many of us wouldn't like them. For instance, mating is seasonal: in season, certain parts of the body turn blue, as with baboons, so there is no more romantic rejection or date rape. And these people can't read, so a lot of harmful ideologies will never trouble them.

There are other genetically engineered creatures in the book as well: Chickie Nobs, for instance, which are chicken objects modified so they grow multiple legs, wings, and breasts. They have no heads, just a nutriment orifice at the top, thus solving a problem for animal rights workers: as their creators say, "No Brain, No Pain." (Since *Oryx and Crake* was published, the Chickie Nob solution has made giant strides: lab-grown meat is now a reality, though it is probably not in your sausages yet.)

A sibling book, *The Year of the Flood,* was published in 2009. Its original title was *God's Gardeners,* but although this was perfectly acceptable to the British publisher, the American publisher and the Canadian publisher objected to it on the grounds that people would think the book was a far-right extremist tract, which goes to show how thoroughly the word *God* has been hijacked. Many other titles were proposed, including "Serpent Wisdom," which the Canadian publisher liked but which the U.S. felt suggested a hippy New Age cult, and "Edencliff," which the British thought sounded like "a retirement home in Bournemouth." Book titles are either immediately obvious, like *The Edible Woman,* or very hard to decide on, and *The Year of the Flood* was the second kind.

The Year of the Flood explores the world of *Oryx and Crake* from a different perspective. Whereas Jimmy/Snowman, the protagonist of *Oryx and Crake,* has grown up within a privileged though barricaded enclave, *The Year of the Flood* takes place in the space outside such enclaves, at the very bottom of the social heap. Its pre-disaster plot unfolds in neighbourhoods that the security forces—now melded with corporations—don't even bother to patrol, leaving them to criminal gangs and anarchic violence. However, this book, too, has a utopia embedded within a dystopia; it's represented by the God's Gardeners, a small environmental religious cult dedicated to the sacred element in all Creation. Its members grow vegetables on slum rooftops, sing sacred-nature hymns, and avoid high-tech communications devices such as cell phones and computers on the grounds that they can be used to spy on you—which is entirely true.

Oryx and Crake and *The Year of the Flood* cover the same time period, and thus are not sequels or prequels; they are more like chapters of the same book. They have sometimes been described as "apocalyptic," but in a true apocalypse everything on Earth is destroyed, whereas in these two books the only element that's annihilated is the human race, or most of it. What survives after the cataclysmic event is not a "dystopia," because many more people would be required for that—enough to comprise a society. The surviving stragglers do, however, have mythic precedents: a number of myths tell of an annihilating flood survived by one man (Deucalion in Greek myth, Utnapishtim in the Gilgamesh epic) or a small group, like Noah and his family. Do the surviving human beings in *Oryx and Crake* and *The Year of the Flood* represent a dystopic threat to the tiny utopia of genetically modified, peaceful, and sexually harmonious New Humans that is set to replace them? As it is always the reader rather than the writer who has the last word about any book, I leave that to you.

. . . . .

People have asked, many times, about the "inspiration" for these two books and their world. Of course there are proximate causes for all novels—a family story, a newspaper clipping, an event in one's personal history—and for *Oryx and Crake* and *The Year of the Flood* there are such causes as well. Worries about the effects of climate change can be found as far back as 1972, when the Club of Rome accurately predicted what now appears to be happening, so those worries had long been with me, though they were not front-page stories in the spring of 2001 when I began *Oryx and Crake*. As with *The Handmaid's Tale*, I accumulated many file folders of research; and although in both there are some of what Huckleberry Finn would call "stretchers," there is nothing that's entirely without foundation.

So I could point to this or that scientific paper, this or that newspaper or magazine story, this or that actual event, but those kinds of things are not really what drive the storytelling impulse. I'm more inclined to think that it's unfinished business, of the kind represented by the questions people are increasingly asking themselves: How badly have we messed up the planet? Can we dig ourselves out? What would a species-wide self-rescue effort look like if played out in actuality? And also: Where has utopian thinking gone? Because it never totally disappears: we're too hopeful a species for that. "Good," for us, may always have a "Bad" twin, but its other twin is "Better."

It's interesting to me that I situated the utopia-facilitating element in *Oryx and Crake* not in a new kind of social organization or a mass brainwashing or soul-engineering program but inside the human body. The Crakers are well behaved from the inside out not because of their legal system or their government or some form of intimidation but because they have been designed to be so. They can't choose otherwise. And this seems to be where ustopia is moving in real life as well: through genetic engineering, we will be able to rid ourselves of inherited diseases, and ugliness, and mental illness, and aging, and . . . who knows? The sky's the limit. Or so we are being told. What is the little dystopia concealed within such utopian visions of the perfected human body—and mind? Time will tell.

Historically, ustopia has not been a happy story. High hopes have been dashed, time and time again. The best intentions have indeed led to many paved roads in Hell. Does that mean we should never try to rectify our mistakes, reverse our disaster-bent courses, clean up our cesspools, or ameliorate the many miseries of many lives? Surely not: if we don't do maintenance work and minor improvements on whatever we actually have, things will go downhill very fast. So of course we should try to make things better, insofar as it lies within our power. But we should probably not try to make things perfect, especially not ourselves, for that path leads to mass graves.

We're stuck with us, imperfect as we are; but we should make the most of us. Which is about as far as I myself am prepared to go, in real life, along the road to ustopia.

NOTES

67 John Gardner, *Grendel*, 1971.

69 *The Tempest*'s Golden Age is in Act II, Scene 1.

69 Underground fairylands and other worlds are numerous. I'll mention only two: George MacDonald's *The Princess and Curdie* (1883) and Aubrey Beardsley's *Under the Hill* (1896).

71 Robert Louis Stevenson wrote an engaging preface to *Treasure Island* in which he describes this process.

71 Escape from caverns via animal tracking is a very old motif. See, for instance, *One Thousand and One Nights.*

73 "Copper cylinders": See James de Mille, *A Strange Manuscript Found in a Copper Cylinder,* 1888.

74 "Hell hath no limits," *Doctor Faustus,* Scene 5, pp. 120–135. "Paradise," *Paradise Lost,* lines 585–7.

76 Honours English: a now-defunct course at the University of Toronto that covered everything from Anglo-Saxon to T. S. Eliot.

78 *At the Back of the North Wind* by George MacDonald (1871) contains a huge flying female character—the North Wind—with an astonishing amount of hair.

79 "Someone once said . . . " Scott Symons, in conversation with the author.

79 Sexual dimorphism in fairy fantasy writing was common in the nineteenth and early twentieth centuries, often featuring a large, long-haired female and a small boy. (See, for instance, Jean Ingelow, *Mopsa the Fairy,* 1910.) In later incarnations such as *She* and *The Lord of the Rings,* power substitutes for size.

80 *Victorian Fairy Painting* is the catalogue of the 1997 exhibit at the Royal Academy of Arts, London.

81 Tennyson and train tracks: "Locksley Hall," 1835.

82 The "city upon a hill" phrase was quoted by George W. Bush, twice, 2000 and 2001.

84 Greg Grandin, *Fordlandia,* 2010; Eduardo Sguiglia, *Fordlandia,* 2000.

88 Emile Zola, *Germinal,* 1885.

Other Deliberations

An Introductory Note

When I began looking through my past publications in search of other pieces about SF and related topics, I found I'd written quite a lot more on this subject than I'd remembered and started quite a bit earlier. My first such published article—on H. Rider Haggard's *She*—dates back to 1965.

I've chosen to begin with an excerpt from a review of Marge Piercy's *Woman on the Edge of Time* that appeared in 1976. It's clear that I was pondering the problems of utopia/dystopia writing at least nine years before I wrote *The Handmaid's Tale.* The other nine pieces range from an introduction to *She* to an examination of Jonathan Swift's science academy in Book III of *Gulliver's Travels* and touch on everything from the forms of dictatorships to human genetic engineering.

I've done some light editing to remove some overlaps and repetitions, but otherwise the pieces are as originally published.

Woman on the Edge of Time by Marge Piercy

None of the reviews of *Woman on the Edge of Time* I've read to date seems even to have acknowledged its genre. Most have assumed that the book is intended as a realistic novel, for that is certainly how it starts out. It appears to be the slice-of-life story of a thirty-seven-year-old Chicana welfare recipient named Consuelo, whose past history we are given in the first few pages of the book. Consuelo had a child, was deserted by her husband, and subsequently took up with a black, blind pickpocket whose death drove her into a depression in which she accidently broke her daughter's wrist. For this offence she was committed to a mental institution and has had her child taken away from her. The only person left for her to love is her doped-up prostitute niece Dolly, but in defending Dolly she breaks the nose of Dolly's pimp and is recommitted by him. The rest of the book takes place "inside" (with one escape and one visit to the outside), and the descriptions of institutional life are enough to make the reader believe that Connie will be driven mad by sadistic doctors and indifferent attendants. This part of the book is rendered in excruciating, grotty, Zolaesque detail, pill by deadening pill, meal by cardboard meal, ordeal by ordeal, and as a rendition of what life in a New York bin is like for those without money or influence it is totally convincing and depressing.

However, even before Connie is recommitted she has been having visits from a strange creature named Luciente. Luciente turns out to be a visitor from the future; Connie thinks the visitor is a young man and is surprised when she is revealed as a woman. By making contact with Connie's mind, Luciente can help Connie project herself into the world of the future, Luciente's world. Connie travels there extensively, and needless to say the reader goes with her.

Some reviewers treated this part of the book as a regrettable daydream or even a hallucination caused by Connie's madness. Such an interpretation undercuts the entire book. If Connie is insane, her struggles to escape from the institution must be viewed in an entirely different light from that in which the author puts them, and the doctors, the pimp, and the indifferent family are somewhat justified in their callous treatment. Other reviewers did not see Connie as insane but took Luciente and her troupe to be a pointless exercise in "science fiction," an exercise that should have no place in a piece of social realism. But Piercy is not that stupid. If she had intended a realistic novel she would have written one. *Woman on the Edge of Time* is a utopia, with all the virtues and shortcomings of the form, and many of the things reviewers found irksome are indigenous to the genre rather than the author.

By utopia, I mean books such as Morris's *News from Nowhere*, Bellamy's *Looking Backward*, Hudson's *A Crystal Age*, or even Wyndham's *Consider Her Ways*. These differ greatly from plot-centred otherworld fantasies such as Tolkien's and though they may share some elements with "science fiction," this category is too broad for them. The books I've mentioned all send an emissary from an oppressive contemporary society into the future as a sort of tourist-journalist, to check out improved conditions and report back. Such books are not really about the hero's adventures, though a love affair of some sort is usually thrown in to sweeten the didactic pill. The real hero is the future society; the reader is intended to comparison-shop in company with the time-traveller, questioning the invariably polite inhabitants and grumbling over disconcerting details. The moral intent of such fables is to point out to us that our own undesirable conditions are not necessary: if things can be imagined differently, they can be done differently.

Hence the inevitable long-winded conversations in which traveller and tour guide, in this case, Connie and Luciente, plod through the day-to-day workings of their societies. What about sewage disposal? birth control? ecology? education? Books of this sort *always* contain conversations like this, and it is to Piercy's credit that she has given us a very human and rather grouchy traveller and a guide who sometimes loses her temper. The world of the future depicted here is closest in spirit perhaps to Morris's. It's a village economy, with each village preserving the ethnic flavour of some worthy present-day minority: American Indian, American Black, European Jewish (suburban WASP is not represented). It is, however, racially mixed, sexually equal, and ecologically balanced. Women have "given up" childbirth in order that men won't regret having given up power, and children are educated more or less communally, with a modified apprentice system. There's quite a lot of advanced bio-feedback, and instant communication through "kenners," which is uncomfortably reminiscent of such silliness as Dick Tracy's two-way wristwatch radios. But they do have communal "fooders" and, I'm happy to note, dishwashers.

Reading utopias is addictive—I found myself skipping through some perfectly acceptable passages about electric shock treatments and visiting hours at the asylum to find out what the inhabitants of Mattapoisett do about breastfeeding (both sexes indulge; men get hormone shots), about motherhood (bottle babies, elective "mothers," production in balance with nature's capacity to support it, adolescent separation rituals), about criminals (if incorrigible they're executed because no one wants to be a prison guard), even about what they use to mulch cabbages. Writing utopias is addictive too, and Piercy expends a good deal of energy trying to get every last detail *in,* to get it *right,* and to make rather too sure we get the point.

Numerous dangers await the author of a utopia. For one thing, inhabitants of utopias somehow cannot help coming across as slightly sanctimonious and preachy; they've been like that since Thomas More. And in addition all utopias suffer from the reader's secret conviction that a perfect world would be dull, so Piercy is careful to liven things up with festivals, ceremonies, nice clothes, and a hopeful description of untrammelled sexual interchange. There are problems, of course, but we are allowed to see the inhabitants working them out through council meetings and "wormings," a wonderful name for a session at which you accuse and complain. Some of these projections are a bit much: it's especially hard to write about communication between cats and humans in any way that isn't whimsical; and utopian children have difficulty being anything but cute or bratty. But the language Piercy has devised for her utopians has unexpected felicities as well as its leaden moments; some of the utopian passages even manage to be oddly moving. The poignancy comes in part from Connie's hunger for human contact and love, in part from the resemblances she sees between the utopians and her lost child, lover, and friends. The outer virtues of Mattapoisett are overshadowed by an inner one: it is the only place where Connie is loved.

However, several issues are dodged. The utopians refuse to fill Connie in on history, so we never find out much about how it all happened. They're engaged in a war with an enemy, but we don't learn much about this either. And they tell Connie they are not "the" future, but only a possible future, and that they need her help in the present to avoid "winking out." (I wish this didn't sound so much like the resuscitation of Tinker Bell in *Peter Pan.*) At one point Connie stumbles into *another* future—presumably what will happen if we don't all put our shoulders to the wheel—in which women are termitelike objects and the air is so polluted you can't see the sky.

The Mattapoisett call to action only bewilders poor Connie, whose scope is of necessity limited. She ends by bumping off a few of the evil asylum shrinks, and because of the ambiguity of the last sections we're left with the uneasy feeling that Mattapoisett may have been a paranoid fantasy after all. The only evidence against this interpretation is that Connie isn't educated enough to have such a utopian vision.

Woman on the Edge of Time is like a long inner dialogue in which Piercy answers her own questions about how a revised American society would work. The curious thing about serious utopias, as opposed to the satirical or entertainment variety, is that their authors never seem to write more than one of them; perhaps because they are products, finally, of the moral rather than the literary sense.

H. Rider Haggard's *She*

When I first read H. Rider Haggard's highly famous novel *She*, I didn't know it was highly famous. I was a teenager, it was the 1950s, and *She* was just one of the many books in the cellar. My father unwittingly shared with Jorge Luis Borges a liking for nineteenth-century yarns with touches of the uncanny coupled with rip-roaring plots; and so, in the cellar, where I was supposed to be doing my homework, I read my way through Rudyard Kipling and Conan Doyle, and *Dracula* and *Frankenstein*, and Robert Louis Stevenson and H. G. Wells, and also Henry Rider Haggard. I read *King Solomon's Mines* first, with its adventures and tunnels and lost treasure, and then *Allan Quatermain*, with its adventures and tunnels and lost civilization. And then I read *She*.

I had no socio-cultural context for these books then—the British Empire was the pink part of the map, "imperialism and colonialism" had not yet acquired their special negative charge, and the accusation "sexist" was far in the future. Nor did I make any distinctions between great literature and any other kind. I just liked reading. Any book that began with some mysterious inscriptions on a very old broken pot was fine with me, and that is how *She* begins. There was even a picture at the front of my edition—not a drawing of the pot but a *photograph* of it, to make the yarn really convincing. (The pot was made to order by Haggard's sister-in-law; he intended it to function like the pirate map at the beginning of *Treasure Island*—a book the popularity of which he hoped to rival—and it did.)

Most outrageous tales state at the very beginning that what follows is so incredible the reader will have trouble believing it, which is both a come-on and a challenge. The messages on the pot stretch credulity, but, having deciphered them, the two heroes of *She*—the gorgeous but none too bright Leo Vincey and the ugly but intelligent Horace Holly—are off to Africa to hunt up the beautiful, undying sorceress who is supposed to have killed Leo's distant ancestor. Curiosity is their driving force, vengeance is their goal. Many a hardship later, and after having narrowly escaped death at the hands of the savage and matrilineal tribe of the Amahaggar, they find not only the ruins of a vast and once-powerful civilization and the numerous mummified bodies of the same but also, dwelling among the tombs, the self-same undying sorceress, ten times lovelier, wiser, and more ruthless than they had dared to imagine.

As Queen of the Amahaggar, "She-who-must-be-obeyed" wafts around wrapped up like a corpse in order to inspire fear; but once tantalizingly peeled, under those gauzy wrappings is a stunner, and—what's more—a virgin. "She," it turns out, is two thousand years old. Her real name is Ayesha. She claims she was once a priestess of the Egyptian nature-goddess Isis. She's been saving herself for two millennia, waiting for the man she loves: one Kallikrates, a very good-looking priest of Isis and the ancestor of Leo Vincey. This man broke his vows and ran off with Leo's ancestress, whereupon Ayesha slew him in a fit of jealous rage. For two thousand years she's been waiting for him to be reincarnated; she's even got his preserved corpse enshrined in a side room, where she laments over it every night. A point-by-point comparison reveals—what a surprise!—that Kallikrates and Leo Vincey are identical.

Having brought Leo to his knees with her knockout charms, and having polished off Ustane, a more normal sort of woman with whom Leo has formed a sexual pair-bond, and who just happens to be a reincarnation of Ayesha's ancient Kallikrates-stealing enemy, Ayesha now demands that Leo accompany her into the depths of a nearby mountain. There, She says, is where the secret of extremely long and more abundant life is to be found. Not only that, She and Leo can't be One until he is as powerful as She—the union might otherwise kill him (as it does, in the sequel, *Ayesha: The Vengeance of She*). So off to the mountain they go, via the ruins of the ancient, once-imperial city of Kôr. To get the renewed life, all one has to do—after the usual Haggard adventures and tunnels—is to traverse some caverns measureless to man, step into a very noisy rolling pillar of fire, and then make one's getaway across a bottomless chasm.

This is how She acquired her powers two thousand years before, and to show a hesitating Leo how easy it is, She does it again. Alas, this time the thing works backward, and in a few instants Ayesha shrivels up into a very elderly bald monkey and then crumbles into dust. Leo and Holly, both hopelessly in love with She and both devastated, totter back to civilization, trusting in Her promise that She will return.

As a good read in the cellar, this was all very satisfactory, despite the overblown way in which She tended to express herself. *She* was an odd book in that it placed a preternaturally powerful woman at the centre of things: the only other such woman I'd run into so far had been the Wonder Woman of the comics, with her sparkly lasso and star-spangled panties. Both Ayesha and Wonder Woman went all weak-kneed when it came to the man they loved—Wonder Woman lost her magic powers when kissed by her boyfriend, Steve Trevor; Ayesha couldn't focus on conquering the world unless Leo Vincey would join her in that dubious enterprise—and I was callow enough, at fifteen, to find this part of it not only soppily romantic but pretty hilarious. Then I graduated from high school and discovered good taste, and forgot for a while about *She*.

.....

For a while, but not forever. In the early 1960s I found myself in graduate school, in Cambridge, Massachusetts. There I was exposed to Widener Library, a much larger and more organized version of the cellar; that is, it contained many sorts of books, not all of which bore the Great Literature Seal of Approval. Once I was let loose in the stacks, my penchant for not doing my homework soon reasserted itself, and it wasn't long before I was snuffling around in Rider Haggard and his ilk once more.

This time, however, I had some excuse. My field of specialization was the nineteenth century, and I was busying myself with Victorian quasi-goddesses; and no one could accuse Haggard of not being Victorian. Like his age, which practically invented archeology, he was an amateur of vanished civilizations; also like his age, he was fascinated by the exploration of unmapped territories and encounters with "undiscovered" native peoples. As an individual, he was such a cookie-cutter country gentleman—albeit with some African travelling in his past—that it was hard to fathom where his overheated imagination had come from, though it may have been this by-the-book-English-establishment quality that allowed him to bypass intellectual analysis completely. He could sink a core-sampling drill straight down into the great English Victorian unconscious, where fears and desires—especially male fears and desires—swarmed in the darkness like blind fish. Or so claimed Henry Miller, among others.

Where did it all come from? In particular, where did the figure of She come from—old-young, powerful-powerless, beautiful-hideous, dweller among tombs, obsessed with an undying love, deeply in touch with the forces of Nature and thus of Life and Death? Haggard and his siblings were said to have been terrorized by an ugly rag doll that lived in a dark cupboard and was named "She-who-must-be-obeyed," but there is more to it than that. *She* was published in 1887, and thus came at the height of the fashion for sinister but seductive women. It looked back also on a long tradition of the same. Ayesha's literary ancestresses include the young-but-old supernatural women in George MacDonald's "Curdie" fantasies, but also various Victorian femmes fatales: Tennyson's Vivien in *The Idylls of the King*, bent on stealing Merlin's magic; the Pre-Raphaelite temptresses created in both poem and picture by Rossetti and William Morris; Swinburne's dominatrixes; Wagner's nasty pieces of female work, including the very old but still toothsome Kundry of *Parsifal*; and, most especially, the Mona Lisa of Walter Pater's famous prose poem, older than the rocks upon which she sits, yet young and lovely, and mysterious, and filled to the brim with experiences of a distinctly suspect nature.

As Sandra Gilbert and Susan Gubar pointed out in their 1989 book, *No Man's Land*, the ascendency in the arts of these potent but dangerous female figures is by no means unconnected with the rise of "Woman" in the nineteenth century, and with the hotly debated issues of her "true nature" and her "rights," and also with the anxieties and fantasies these controversies generated. If women ever came to wield political power—to which they were surely, by their natures, unsuited—what would they do with it? And if they were beautiful and desirable women, capable of attacking on the sexual as well as the political front, wouldn't they drink men's blood, sap their vitality, and reduce them to grovelling serfs? As the century opened, Wordsworth's Mother Nature was benign, and "never would betray/The heart that loved her"; but by the end of the century, Nature and the women so firmly linked to her were much more likely to be red in tooth and claw—Darwinian goddesses rather than Wordsworthian ones. When, in *She*, Ayesha appropriates the fiery phallic pillar at the heart of Nature for the second time, it's just as well that it works backward. Otherwise men could kiss their own phallic pillars goodbye.

"You are a whale at parables and allegories and one thing reflecting another," wrote Rudyard Kipling in a letter to Rider Haggard, and there appear to be various hints and verbal signposts scattered over the landscape of *She*. For instance, the Amahagger, the tribe ruled by She, bear a name that not only encapsulates *hag* but also conflates the Latin root for *love* with the name of Abraham's banished wilderness-dwelling concubine, Hagar, and thus brings to mind a story of two women competing for one man. The ancient city of Kôr is named perhaps for *core*, cognate with the French *coeur*, but suggesting also *corps*, for body, and thus *corpse*, for dead body; for She is in part a Nightmare Life-in-Death. Her horrid end is reminiscent of Darwinian evolution played backward—woman into monkey—but also of vampires after the stake-into-the-heart manoeuvre. (Bram Stoker's *Dracula* appeared after *She*, but Sheridan LeFanu's *Carmilla* pre-dates it, as does many another vampire story.) These associations and more point toward some central significance that Haggard himself could never fully explicate, though he chalked up a sequel and a couple of prequels trying. "*She*," he said, was "some gigantic allegory of which I could not catch the meaning."

Haggard claimed to have written *She* "at white heat," in six weeks—"It came," he said, "faster than my poor aching hand could set it down," which would suggest hypnotic trance or possession. In the heyday of Freudian and Jungian analysis, *She* was much explored and admired, by Freudians for its womb-and-phallus images, by Jungians for its anima figures and thresholds. Northrop Frye, proponent of the theory of archetypes in literature, says this of *She* in his 1975 book, *The Secular Scripture: A Study of the Structure of Romance*:

> In the theme of the apparently dead and buried heroine who comes to life again, one of the themes of Shakespeare's *Cymbeline*, we seem to be getting a more undisplaced glimpse of the earth-mother at the bottom of the world. In later romance there is another glimpse of such a figure in Rider Haggard's *She*, a beautiful and sinister female ruler, buried in the depths of a dark continent, who is much involved with archetypes of death and rebirth. . . . Embalmed mummies suggest Egypt, which is preeminently the land of death and burial, and, largely because of its biblical role, of descent to a lower world.

Whatever *She* may have been thought to signify, its impact upon publication was tremendous. *Everyone* read it, especially men; a whole generation was influenced by it, and the generation after that. A dozen or so films have been based on it, and a huge amount of the pulp-magazine fiction churned out in the teens, twenties, and thirties of the twentieth century bears its impress. Every time a young but possibly old and/or dead woman turns up, especially if she's ruling a lost tribe in a wilderness and is a hypnotic seductress, you're looking at a descendant of She.

Literary writers, too, felt Her foot on their necks. Conrad's *Heart of Darkness* owes a lot to Her, as Gilbert and Gubar have indicated. James Hilton's Shangri-La, with its ancient, beautiful, and eventually crumbling heroine, is an obvious relative. C. S. Lewis felt Her power, fond as he was of creating sweet-talking, good-looking evil queens; and in Tolkien's *The Lord of the Rings*, She splits into two: Galadriel, powerful but good, who's got exactly the same water-mirror as the one possessed by She; and a very ancient cave-dwelling man-devouring spider-creature named, tellingly, Shelob.

Would it be out of the question to connect the destructive Female Will, so feared by D. H. Lawrence and others, with the malign aspect of She? For Ayesha is a supremely transgressive female who challenges male power; though Her shoe size is tiny and Her fingernails are pink, She's a rebel at heart. If only She hadn't been hobbled by love, She would have used her formidable energies to overthrow the established civilized order. That the established civilized order was white and male and European goes without saying; thus She's power was not only female—of the heart, of the body—but barbaric, and "dark."

By the time we find John Mortimer's Rumpole of the Bailey referring to his dumpy, kitchen-cleanser-conscious wife as "she who must be obeyed," the once-potent figure has been secularized and demythologized, and has dwindled into the combination of joke and rag doll that it may have been in its origins. Nevertheless, we must not forget one of Ayesha's pre-eminent powers—the ability to reincarnate herself. Like the vampire dust at the end of Christopher Lee movies, blowing away only to reassemble itself at the outset of the next film, She could come back. And back. And back.

No doubt this is because She is in some ways a permanent feature of the human imagination. She's one of the giants of the nursery, a threatening but compelling figure, bigger and better than life. Also worse, of course. And therein lies her attraction.

NOTES

99 Margaret Atwood, "Superwoman Drawn and Quartered: The Early Forms of *She*," *Alphabet* magazine, vol. 10, July 1965.

112 Northrop Frye, *The Secular Scripture: A Study of the Structure of Romance* (Cambridge, Mass.: Harvard University Press, 1976).

113 Sandra M. Gilbert and Susan Gubar, *No Man's Land: The Place of the Woman Writer in the Twentieth Century, vol. 2, Sexchanges* (New Haven: Yale University Press, 1989).

113 Daniel Karlin, Introduction, in H. Rider Haggard, *She* (Oxford: Oxford University Press, 1991).

The Queen of Quinkdom: *The Birthday of the World and Other Stories* by Ursula K. Le Guin

The Birthday of the World is Ursula K. Le Guin's tenth collection of stories. In it she demonstrates once again why she is the reigning queen of . . . but immediately we come to a difficulty, for what is the fitting name of her kingdom? Or, in view of her abiding concern with the ambiguities of gender, her queendom, or perhaps—considering how she likes to mix and match—her quinkdom? Or may she more properly be said to have not one such realm, but two?

"Science fiction" is the box in which her work is usually placed, but it's an awkward box: it bulges with discards from elsewhere. Into it have been crammed all those stories that don't fit comfortably into the family room of the socially realistic novel or the more formal parlour of historical fiction, or other compartmentalized genres: westerns, gothics, horrors, gothic romances, and the novels of war, crime, and spies. Its subdivisions include science fiction proper (gizmo-riddled and theory-based space travel, time travel, or cybertravel to other worlds, with aliens frequent); science-fiction fantasy (dragons are common; the gizmos are less plausible, and may include wands); and speculative fiction (human society and its possible future forms, which are either much better than what we have now or much worse). However, the membranes separating these subdivisions are permeable, and osmotic flow from one to another is the norm.

.....

The lineage of "science fiction," broadly considered, is very long, and some of its literary ancestors are of the utmost respectability. Alberto Manguel has catalogued many in *The Dictionary of Imaginary Places*: Plato's account of Atlantis is among them, and Sir Thomas More's *Utopia* and Swift's *Gulliver's Travels*. Accounts of voyages to unknown realms with bizarre inhabitants are as old as Herodotus in his wilder moments, as old as *One Thousand and One Nights*, as old as Thomas the Rhymer. Folk tales, the Norse Sagas, and the adventure romances of chivalry are not-so-distant cousins of such tales, and have been drawn on by hundreds of imitators of *The Lord of the Rings* and/or *Conan the Conqueror*—works that previously fetched their water from the same wells, as did their precursors, George MacDonald and the H. Rider Haggard of *She*.

Jules Verne is probably the best known of the early gizmo-fictionalists, but Mary Shelley's *Frankenstein* could be thought of as the first "science fiction"—that is, the first fiction that had real science in it—inspired as it was by experiments with electricity, in particular the galvanizing of corpses. Some of her preoccupations have stayed with the genre (or genres) ever since: most specifically, what is the price that must be paid by Promethean Man for stealing fire from Heaven? Indeed, some commentators have proposed "science fiction" as the last fictional repository for theological speculation. Heaven, Hell, and aerial transport by means of wings having been more or less abandoned after Milton, outer space was the only remaining neighbourhood where beings resembling gods, angels, and demons might still be found. J. R. R. Tolkien's friend and fellow fantasist C. S. Lewis even went so far as to compose a "science fiction" trilogy—very light on science but heavy on theology, the "spaceship" being a coffin filled with roses and the temptation of Eve being re-enacted on the planet of Venus, complete with luscious fruit.

Rearranged human societies have been a constant in the tradition as well, and they have been used both to criticize our present state of affairs and to suggest more pleasant alternatives. . . . The nineteenth century, cheered on by its successes with sewage systems and prison reform, produced so many earnestly hopeful speculative fictions that the vogue was satirized not only by Gilbert and Sullivan's operetta *Utopia Limited* but also by Samuel Butler's *Erewhon*, where illness is a crime and crime is an illness.

However, as the optimism of the nineteenth century gave way to the Procrustean social dislocations of the twentieth—most notably in the former Soviet Union and the former Third Reich—literary utopias, whether serious or sardonic, were displaced by darker versions of themselves. . . . *Brave New World* and *Nineteen Eighty-Four* are of course the best known of these many prescient badlands, with Karel Čapek's *R.U.R.* and the nightmarish fables of John Wyndham running close behind.

It's too bad that one term—*science fiction*—has served for so many variants, and too bad also that this term has acquired a dubious if not downright sluttish reputation. True, the proliferation of sci-fi in the 1920s and 1930s gave rise to a great many bug-eyed-monster-bestrewn space operas, followed by films and television shows that drew heavily on this odiferous cache. . . .

In brilliant hands, however, the form can be brilliant. . . .

Which brings us to Ursula K. Le Guin. No question about her literary quality: her graceful prose, carefully thought-through premises, psychological insight, and intelligent perception have earned her the National Book Award, the Kafka Award, five

Hugos, five Nebulas, a Newbery, a Jupiter, a Gandalf, and an armful of other awards, great and small. Her first two books, *Planet of Exile* and *Rocannon's World,* were published in 1966, and since then she has published sixteen novels, as well as ten collections of stories.

Collectively, these books have created two major parallel universes: the universe of the Ekumen, which is sci-fi proper—spaceships, travel among worlds, and so forth—and the world of Earthsea. The latter must be called "fantasy," I suppose, as it contains dragons and witches and even a school for wizards, though this institution is a long way from the Hogwarts of Harry Potter. The Ekumen series may be said—very broadly—to concern itself with the nature of human nature: How far can we stretch and still remain human? What is essential to our being, what is contingent? The Earthsea series is occupied—again, very broadly speaking—with the nature of reality and the necessity of mortality, and also with language in relation to its matrix. (That's heavy weather to make of a series that has been promoted as suitable for age twelve, but perhaps the fault lies in the marketing directors. Like *Alice In Wonderland*, these tales speak to readers on many levels.)

Le Guin's preoccupations are not divided into two strictly separate packages, of course: both of her worlds are scrupulously attentive to the uses and misuses of language; both have their characters fret over social gaffes and get snarled up in foreign customs; both worry about death. But in the Ekumen universe, although there is much strangeness, there is no magic, apart from the magic inherent in creation itself.

The astonishing thing about Le Guin as a writer is that she managed to create these two realms, not only in parallel, but at the same time. The first Earthsea book, *A Wizard of Earthsea*, appeared in 1968, and *The Left Hand of Darkness*, the famous classic from the Ekumen series, in 1969. Either one would have been sufficient to establish Le Guin's reputation as a mistress of its genre; both together make one suspect arcane drugs or creative double-jointedness or ambidexterity. Not for nothing did Le Guin invoke handedness in her fourth title: as soon as we start talking about the *left* hand, all sorts of biblical connotations gather. (Although the left hand is the sinister one, God too has a left hand, so left hands can't be all bad. Should your right hand know what your left hand is doing, and if not, why not? And so forth.) As Walter Benjamin once said, the decisive blows are struck left-handed.

Ursula K. Le Guin has continued to explore and describe and dramatize both of her major fictional realms over the thirty-six years that have passed since her first novel was published. But since the stories in *The Birthday of the World* are Ekumen stories—with two exceptions—it's as well to concentrate on the science-fiction world rather than on the fantasy one. The general premises of the Ekumen series are as follows. There are many habitable planets in the universe. Long, long ago they were "seeded" by a people called the Hainish, space travellers from an Earthlike planet, after which time passed, disruptions occurred, and each society was left alone to develop along different lines.

Now, a benevolent federation called the Ekumen having been established, explorers are being sent out to see what has become of these far-flung but still hominid or perhaps even human societies. Conquest is not the aim, nor is missionary work: non-invasive, non-directive understanding and recording are the functions required of such explorers or ambassadors, who are known as Mobiles. Various gizmos are provided to allow them to function amid the alien corn, and they are provided with a handy widget called the "ansible," a piece of technology we should all have because it allows for instantaneous transmission of information, thus cancelling out the delaying effects of the fourth dimension. Also, it never seems to crash like your Internet e-mail program. I'm all for it.

. . . . .

Here it is necessary to mention that Le Guin's mother was a writer, her husband is an historian, and her father was an anthropologist; thus she has been surrounded all her life by people whose interests have dovetailed with her own. The writing connection, through her mother, is obvious. Her husband's historical knowledge must have come in very handy: there's more than an echo in her work of the kinds of usually unpleasant events that change what we call "history." But her father's discipline, anthropology, deserves special mention.

If the "fantasy" end of science fiction owes a large debt to folk tale and myth and saga, the "science fiction" end owes an equally large debt to the development of archeology and anthropology as serious disciplines, as distinct from the tomb-looting and exploration-for-exploitation that preceded them and continued alongside them. Layard's discovery of Nineveh in the 1840s had the effect of a can opener on Victorian thinking about the past; Troy and Pompeii and ancient Egypt were similarly mesmerizing. Through new discoveries and fresh excavations, European concepts of past civilizations were rearranged, imaginative doors were opened, wardrobe choices were expanded. If things were once otherwise, perhaps they could be otherwise again, especially where clothing and sex were concerned—two matters that particularly fascinated Victorian and early twentieth-century imaginative writers, who longed for less of the former and more of the latter.

Anthropology arrived a little later. Cultures were discovered in remote places that were very different from the modern West, and rather than being wiped out or subjugated, they were taken seriously and studied. How are these people like us? How are they different? Is it possible to understand them? What are their foundation myths, their beliefs about an afterlife? How do they arrange their marriages, how do their kinship systems work? What are their foods? How about their (a) clothing and (b) sex? Which were usually discovered—through the work of various perhaps overeager inquirers such as Margaret Mead—to be (a) scantier and (b) more satisfactory than ours.

Anthropologists do—or are supposed to do—more or less what the Mobiles in Le Guin's Ekumen construction are supposed to do: they go to distant shores, they look, they explore foreign societies and try to figure them out. Then they record, and then they transmit. Le Guin knows the tricks of the trade, and also the pitfalls: her Mobiles are mistrusted and misled while they are in the field, just as real anthropologists have been. They're used as political pawns, they're scorned as outsiders, they're feared because they have unknown powers. But they are also dedicated professionals and trained observers, and human beings with personal lives of their own. This is what makes them and the stories they tell believable, and Le Guin's handling of them engaging as writing in its own right.

. . . . .

It's informative to compare two of Le Guin's introductions: the one she wrote for *The Left Hand of Darkness* in 1976, seven years after the book was first published, and the foreword she's now written for *The Birthday of the World. The Left Hand of Darkness* takes place on the planet of Gethen, or Winter, where the inhabitants are neither men nor women nor hermaphrodites. Instead they have phases: a non-sexual phase is followed by a sexual phase, and during the latter each individual changes into whichever gender is suitable for the occasion. Thus anyone at all may be, over a lifetime, both mother and father, both penetrator and penetree. As the story opens, the "king" is both mad and pregnant, and the non-Gethenian observer from the Ekumen is nothing if not confused.

This novel appeared at the beginning of the hottest period of 1970s feminism, when emotions were running very high on subjects having to do with genders and their roles. Le Guin was accused of wanting everyone to be an androgyne and of predicting that in the future they would be; conversely, of being anti-feminist because she'd used the pronoun "he" to denote persons not in "kemmer"—the sexual phase.

Her introduction to *The Left Hand of Darkness* is therefore somewhat brisk. Science fiction should not be merely extrapolative, she says; it should not take a present trend and project it into the future, thus arriving via logic at a prophetic truth. Science fiction cannot predict, nor can any fiction, the variables being too many. Her own book is a "thought-experiment," like *Frankenstein.* It begins with "Let's say," follows that with a premise, and then watches to see what happens next. "In a story so conceived," she says, "the moral complexity proper to the modern novel need not be sacrificed . . . thought and intuition can move freely within bounds set only by the terms of the experiment, which may be very large indeed."

The purpose of a thought-experiment, she writes, is to "describe reality, the present world." "A novelist's business is lying"—lying interpreted in the novelist's usual way, that is, as a devious method of truth-telling. Consequently the androgyny described in her book is neither prediction nor prescription, just description: androgyny, metaphorically speaking, is a feature of all human beings. With those who don't understand that metaphor is metaphor and fiction is fiction, she is more than a little irritated. One suspects she's received a lot of extremely odd fan mail.

The Foreword to *The Birthday of the World* is mellower. Twenty-six years later, the author has fought her battles and is an established feature of the sci-fi landscape. She can afford to be less didactic, more charmingly candid, a little scattier. The universe of the Ekumen now feels comfortable to her, like "an old shirt." No sense in expecting it to be consistent, though: "Its Time Line is like something a kitten pulled out of the knitting basket, and its history consists largely of gaps." In this Foreword, Le Guin describes process rather than theory: the genesis of each story, the problems she had to think her way through. Typically, she doesn't concoct her worlds: she finds herself in them, and then begins to explore them, just like, well, an anthropologist. "First to create difference," she says, ". . . then to let the fiery arc of human emotion leap and close the gap: this acrobatics of the imagination fascinates me and satisfies me as almost no other."

. . . . .

There are seven shorter stories in *The Birthday of the World*, and one that might qualify as a novella. Six of the first seven are Ekumen stories—they're part of the "old shirt." The seventh probably belongs there, though its author isn't sure. The eighth is set in a different universe altogether—the generic, shared, science-fiction "future." All but the eighth are largely concerned with—as Le Guin says—"peculiar arrangements of gender and sexuality."

All imagined worlds must make some provision for sex, with or without black leather and tentacles, and the peculiarity of the arrangements is an old motif in science fiction: one thinks not only of Charlotte Perkins Gilman's *Herland*, where the genders live separately, but also of W. H. Hudson's *A Crystal Age*, featuring an antlike neuter state, or John Wyndham's *Consider Her Ways*, also based on a hymenoptera model, or Marge Piercy's *Woman on the Edge of Time*, which tries for absolute gender equality. (Men breast-feed: watch for this trend.) But Le Guin takes things much further. In the first story, "Coming of Age in Karhide," we see Gethen/Winter not through the eyes of a Mobile but through those of a Gethenian just coming into adolescence: which gender will s/he turn into first? This story is not only erotic but happy. Why not, in a world where sex is always either spectacular or of no concern whatsoever?

Things aren't so jolly in "The Matter of Seggri," where there's a gender imbalance: far more women than men. The women run everything, and marry each other as life partners. The rare boy children are spoiled by the women, but as men they must live a segregated life in castles, where they dress up, show off, stage public fights, and are rented out as studs. They don't have much fun. It's like being trapped in the World Wrestling Federation, forever.

"Unchosen Love" and "Mountain Ways" take place on a

world called O, created by Le Guin in *A Fisherman of the Inland Sea.* On O, you must be married to three other people but can have sex with only two of them. The quartets must consist of a Morning man and a Morning woman—who can't have sex—and an Evening man and an Evening woman, who also can't have sex. But the Morning man is expected to have sex with the Evening woman and also the Evening man, and the Evening woman is expected to have sex with the Morning man and also the Morning woman. Putting these quartets together is one of the problems the characters face, and keeping them straight—who's for you, who's taboo—is a problem for both reader and writer. Le Guin had to draw charts. As she says, "I like thinking about complex social relationships which produce and frustrate highly charged emotional relationships."

"Solitude" is a meditative story about a world in which conviviality is deeply distrusted. Women live alone in their own houses in an "auntring" or village, where they make baskets and do gardening, and practise the non-verbal art of "being aware." Only the children go from house to house, learning lore. When girls come of age they form part of an auntring, but boys must go off to join adolescent packs and scratch a living in the wilderness. They fight it out, and those who survive become breeding males, living shyly in hermit huts, guarding the auntrings from a distance, and being visited by the women, who "scout" for purposes of mating. This setup, despite its spiritual satisfactions, would not suit everyone.

. . . . .

"Old Music and the Slave Women" comes very close to home, inspired as it was by a visit to a former plantation in the American

South. On the planet of Werel, slavers and anti-slavers are at war, and sex among the slavers is a matter of raping the field hands. The chief character, an intelligence officer with the Ekumen embassy, gets into arguments over human rights and then bad trouble. Of all the stories, this one comes closest to substantiating Le Guin's claim that science fiction describes our own world. Werel could be any society torn by civil war: wherever it's happening, it's always brutal, and Le Guin, although at times a movingly lyrical writer, has never shied away from necessary gore.

The title story is constructed on an Inca base, with a splash of ancient Egypt. A man and a woman together form God. Both positions are hereditary and created by brother–sister marriage; the duties of God include divination by dancing, which causes the world to be born anew each year. Governance is carried out by God's messengers, or "angels." What happens when a foreign but powerful presence enters this highly structured world and the belief system that sustains it crumbles? You can imagine, or you can read *The Conquest of Peru.* Nevertheless, this delicate story is strangely courageous, strangely hopeful: the world ends, but then, too, it is always beginning.

The last story, "Paradises Lost," continues the note of renewal. Many generations have been born and have died onboard a long-distance spaceship. During the voyage a new religion has sprung up, whose adherents believe they are actually, now, in Heaven. (If so, Heaven is just as boring as some have always feared.) Then the ship reaches the destination proposed for it centuries earlier, and its inhabitants must decide whether to remain in "Heaven" or to descend to a "dirtball" whose flora, fauna, and microbes are completely alien to them. The most enjoyable part of this story, for me, was the release from claustrophobia: try as I might, I couldn't imagine why anyone would prefer the ship.

Le Guin is on the side of the dirtball too; and, by extension, of our very own dirtball. Whatever else she may do—wherever her curious intelligence may take her, whatever twists and knots of motive and plot and genitalia she may invent—she never loses touch with her reverence for the immense *what is*. All her stories are, as she has said, metaphors for the one human story; all her fantastic planets are this one, however disguised. "Paradises Lost" shows us our own natural world as a freshly discovered Paradise Regained, a realm of wonder; and in this, Le Guin is a quintessentially American writer, of the sort for whom the quest for the Peaceable Kingdom is ongoing. Perhaps, as Jesus hinted, the kingdom of God is within; or perhaps, as William Blake glossed, it is within a wildflower, seen aright.

The story—and the book—ends with a minimalist dance, as an old woman and a crippled old man celebrate, indeed worship, the ordinary dirt that sustains them. "Swaying, she lifted her bare feet from the dirt and set them down again while he stood still, holding her hands. They danced together that way."

Arguing Against Ice Cream: *Enough: Staying Human in an Engineered Age* by Bill McKibben

Enough, by Bill McKibben, is a passionate, succinct, chilling, closely argued, sometimes hilarious, touchingly well-intentioned, and essential summary of the future proposed by "science" for the human race. This is the same Bill McKibben who wrote *The End of Nature*, about how *Homo sapiens* has been rearranging the biosphere with the aid of genetically modified plants to suit what it believes is its own interests, and *Long Distance*, about running marathons, and many essays for *The New Yorker*, the *New York Times*, *The New York Review of Books*, *The Atlantic*, and others.

Bill McKibben appears to be a smart and thoughtful person but also kindly and optimistic, as far as can be told from his prose. He likes going for walks in the woods, and he seems very fit, and his jacket photo looks like someone you wouldn't want playing against you at bridge because he'd already know what you had in your hand. In other words, he could qualify for membership in a muscular branch of upper-level-I.Q. geekhood, and cannot be simply dismissed as a dull-normal Luddite too dumb to understand the nifty customized body-and-brain parts soon to be on offer to you and yours.

On offer for a price, of course. Ah yes, the price. The traditional fee for this kind of thing was your soul, but who pays any attention to that tattered theological rag anymore, since it can't be located with a brain probe? And hey, the Special Deal is a super package! How could you refuse? It contains so much that human dreams are made of.

Faust wanted the same sort of stuff. Many have wanted it: eternal youth, godlike beauty, hyper-intelligence, Charles Atlas strength. Those of us brought up on the back pages of comic books know the appeal. They'll never laugh again when you sit down at the piano because now you'll have X-Men fingers and Mozart's genius; they won't dare to kick sand in your face at the beach because you'll be built like Hercules; you'll never again be refused a date because of your ugly blackheads, which will have been banished, along with many another feature you could do without. Turning to more adult concerns such as death, you won't have to invest in a cement coffin container because not only will your loved one be safe tonight, but he or she will still be alive, and forever! And so will you.

The line forms to the right, and it'll be a long one. (*Enough* mentions a couple of California artists who set up a piece of conceptual art in the form of a boutique called Gene Genies Worldwide, with printed brochures illustrating what you could buy, and were deluged with serious inquiries.) Anyone who thinks there won't be a demand for what's putatively on sale is hallucinating. But along comes Bill McKibben with his sidewalk-preacher's sandwich board, denouncing the whole enterprise and prophesying doom. There will be catcalls of *killjoy* and *spoilsport*, not to mention *troglodyte*, *nay-sayer*, and *hand-wringer*. Like Prince Charles, who's just come out against nanotechnology on the grounds that it could reduce the world to grey goo, McKibben will be told to keep his nose out of it because it's none of his business.

"Mankind was my business," laments Marley's ghost when it's too late for him. And so says Bill McKibben. Mankind is his business. He addresses the greedy little Scrooge in all of us and points out to that greedy little Scrooge why he should not want more and more, and more, and, just to top it off, more.

More of what? To that in a minute, but first, a digression on the word *more.* Two emblematic uses of *more* spring to mind. The first is, of course, the echoing "more" pronounced by Oliver Twist when he is being starved in a foundlings' home by venal officials. That "more" is the legitimate response to "not enough." It's the "more" of real need, and only the hard-hearted and wickedly self-righteous Mr. Bumbles of this world can be outraged by it. The second "more" is in the film *Key Largo,* in the remarkable exchange between the Humphrey Bogart hero character and the Edward G. Robinson evil crook. The crook is asked what he wants, and he doesn't know. Humphrey knows, however. "He wants more," he says. And this is what the crook does want: more, and more than he can possibly use; or, rather, more than he can appreciate, dedicated as he is to mere accumulation and mere power. For the alternative to "more," in McKibben's book, is not "less" but "enough." Its epigraph might well be that old folk saying, "Enough is as good as a feast."

The "enough" of the title, seen rightly—McKibben implies—is already a feast. It's us, as we are, with maybe a few allowable improvements. More than that is too much. These tempting "mores"—for there are many of them—grow on the more and more Trees of Knowledge that crowd the modern scientific landscape so thickly you can't see the forest for them. McKibben takes axe in hand and sets out to clear a path. Which apples should be plucked, which left alone? How hard should we think before taking the fateful bite? And why shouldn't we pig out, and what's our motivation if we do? Is it the same old story—we want to be as the gods? If it's *that* story, we've read it, in its many versions. It's never had a happy ending. Not so far.

. . . . .

The items on the smorgasbord of human alteration divide roughly into three. First, genetic alteration, or gene splicing, whereby parents who are five feet tall and bald can give birth to a six-footer with long blond hair who looks like the next-door neighbour. Well, it'll provide some new excuses. ("Honey, we *chose* that! Remember?") Second, nanotechnology, or the development of single-atom-layer gizmos that can replicate themselves and assemble and disassemble matter. Some of these might be sent into our bodies to repair them, like the miniaturized submarine containing the memorable Raquel Welch in the film *Fantastic Voyage*. Third, cybernetics, or the melding of man with machine, like the bionic man. At least we'll all be able to get the lids off jars.

There's a fourth idea that's glanced at—cryogenics, or getting yourself or your budget-version head flash-frozen until such time as the yellow-brick road to immortality has been built; whereupon you'll be unfrozen and restored to youth and health, and, if the head-only option has been chosen, a new body can be grown for you from a few scrapings of your—or somebody else's—DNA. Investing even a small amount of belief in this scheme puts you in the same league as those who happily buy the Brooklyn Bridge from shifty-looking men in overcoats, for the company—yes, it would be a company—in charge of your frozen head would need to be not only perennially solvent—bankruptcy would equal meltdown—but also impeccably honest.

Every field of human endeavour attracts its quota of con men and scam artists, but this one would seem to be a natural. What's to stop the operators from banking your money, subjecting you to the initial gelatification, and then, pleading electrical failure, dumping your unpleasantly melting self into the trash, or, better—waste not, want not, and the shareholders expect a solid bottom line—recycling you for cat food? The Pyramids of the mummified Egyptian kings, thoroughly pillaged once the relatives' backs were turned, stand as a gloss on this kind of thinking, as does London's Highgate Cemetery, a Garden of Eternity parcelled out in pricey lots that became an overgrown thicket once the money stream petered out.

But McKibben's fervent arguments are of a more clean-cut kind: he is not a novelist or a poet, and thus does not descend all the way into the foul rag-and-bone shop of the heart. He assumes a certain amount of sincerity and probity in the less-wacky advocates of these developments, and his appeals are directed to our rational and ethical faculties. We should act, he believes, out of respect for human history and the human race.

He first tackles genetic engineering, already present in soybeans and not so far off for *Homo sapiens* now that we have the luminous green rabbit and the goat/spider. Gene splicing is the modern answer to the eternal urge to make a more perfect model of ourselves. The novel of record is Mary Shelley's *Frankenstein*: we just can't stop tinkering, partly because it's so interesting and partly because we have a high opinion of our own abilities; but we risk creating monsters.

Gene splicing depends on cloning—McKibben explains how—but is not the same. It involves inserting selected genes—of those other than the parents—into an egg, which is then implanted in the usual way (or will be until the bottled babies of *Brave New World* make their appearance and we can do away with the womb altogether). If we become genetically enhanced in this way—enhanced by our parents before we're born—the joy and mystery will go out of life, says McKibben, because we won't have to strive for mastery. Our achievements won't be "ours" but will have been programmed into us; we'll never know whether we are really feeling "our" emotions, or whether they—like the false memories embedded in the replicants in the film *Blade Runner*—are off the shelf. We won't be our unique selves, we'll just be the sum totals of market whims. We truly will be the "meat machines" that some scientists already term us. Right now about all our parents can pick for us are our names, but what if they could pick everything about us? (And you thought your mother had bad taste in sofas!)

Worse, we'll be caught in a keep-up-with-the-Joneses competition whereby each new generation of babies will have to have all the latest enhancements—will have to be more intelligent, more beautiful, more disease-free, longer-lived, than the generation before. (Babies of the rich, it goes without saying, because there's gold in them thar frills.) Thus each new generation will be *sui generis*—isolated, disconsolate, as out of date as last year's car model before they're even twenty-one, each of them stuck on a lily pad of enhancement a few hops behind the one that follows them. In addition to that, they'll be cut off from history—from their own family tree—because who knows what family trees they'll really be perpetuating? They'll bear little relation to their so-called ancestors. The loneliness and the sense of disconnection could be extreme.

. . . . .

McKibben does not go on to explore the ultimate hell this situation could produce. Imagine the adolescent whining and sulking that will be visited upon the parents who have chosen their children's features out of a catalogue and—inevitably—will have chosen wrong. "I didn't ask to be born" will be replaced by resentments such as "I didn't ask to have blue eyes" or "I didn't ask to be a math whiz." Burn that gene brochure! If your kid whines about not being enhanced enough, you can just say you couldn't afford it. (The advocates of gene enhancement might respond by saying that since you'll be able to choose your child's temperament as well, naturally you'll pick a type that will never do any adolescent whining or sulking. Pay no attention: these people will not be talking about flesh-and-blood children, but about Stepford Kids.)

Again, McKibben doesn't go all the way down, into the dark realms of envy, cheating, payoffs, and megalomaniacal revenge. What's to prevent your enemy from bribing your gene doctor so that your baby turns out like Hannibal the Cannibal?

But what about heritable diseases? you may reasonably ask. Why should any child get stuck with cerebral palsy, or autism, or schizophrenia, or Huntington's chorea, or the many other maladies that genes are heir to? They shouldn't if there's a remedy, and there is. McKibben points out that these conditions can be eliminated without taking the final step. (After *Enough* was published and before this review was written, a Canadian team cracked the gene for autism. Help is on the way.) Once their genome has been analyzed, parents at risk could be notified of any defects, and could go the *in vitro* route, with fertilized eggs lacking the culpable gene chosen for implanting. This "somatic gene therapy" would not involve the addition of anyone else's genes. Plastic surgery, hormones, vitamin pills, and somatic gene therapy are enough, says McKibben; gene splicing is too much.

Next, McKibben delves into nanotechnology, which is also well on the way. The applicable folk tale for nanotechnology is "The Sorcerer's Apprentice"—what if you get the process started, but the self-replicating nanobot escapes, and you can't turn the darn thing off? We might create an assembler that makes food—dirt in one end, potatoes out the other—or something that destroys bioforms hostile to us. But what if such a nanobot goes on the rampage and attacks all bioforms? This is where Prince Charles's apprehension about "grey goo" comes in. It's a real fear, and one discussed by McKibben.

Cybernetics and artificial intelligence also get a look-in, as man-and-machine combinations are occupying some of our better-paid minds. Visions of microchips implanted in your brain dance in their heads—well, we already have pacemakers, so what's the difference? Why shouldn't we baptize artificial intelligence doodads because they can be made to resemble us so much that maybe they have whatever we think merits baptism? Call it a soul; why not? Maybe we can get enhanced smellability, X-ray vision, Spidey Sense, the works. Artificial orgasms, better than the real thing. *Everything* will be better than the real thing! Why shouldn't we have eyes in the backs of our heads? Why do we only have one mouth that has to perform several functions—talking, eating, whistling? If we had several buccal orifices we'd be able to do all these things at once! (Sign here. You owe it to yourself. Because you deserve it.)

There's been quite a lot of chat about the shortcomings we've had to put up with due to Mother Nature, the dirty, treacherous cow, and this is the not-so-cleverly-hidden subtext of a lot of *Brave New World*–type thinking. These folks hate Nature, and they hate themselves as part of it, or her. McKibben cites an amazing speech given by Max More (last name chosen by himself) to the Extropian Convention ("extropy," coined as the opposite to "entropy"). This speech took the form of a dissing of Mother Nature, and said, essentially, thanks for nothing and goodbye. Nature has made so many mistakes, the chief one being death. Why do we have to get old and die? Why is man the one creature that foresees his own death?

As in many religions—and the energy propelling the wilder fringes of this "more" enterprise is religious in essence—there has to be a second birth, one that gets around the indignity of having come out of a body—a female body—and, come to think of it, of having a body yourself. All that guck and blood and cells and death. Why do we have to eat? And, by implication, defecate. So messy. Maybe we can fix our digestive tracts so we just slip out a little pellet—say, once a month. Maybe we can be born again, this time out of an artificial head instead of a natural body, and download the contents of our brains into machines, and linger around in cyberspace, as in William Gibson's novels. Though if you've read William Gibson, you'll know the place is a queasy nightmare.

All the enhancements McKibben discusses are converging on the biggie, which is none other than the final nose-thumb at Nature—immortality. Immortality doesn't fare so well in myth and story. Either you get it but forget to request eternal youth too and become a crumbling horror (Tithonus, the Sibyl of Cumae, Swift's Struldbrugs), or you seize the immortality and the vitality but lose your soul and must live by feeding on the blood of the innocent (Melmoth the Wanderer, vampires, and so forth). The stories are clear: gods are immortal, men die, try to change it and you'll end up worse off.

That doesn't stop us from hankering. McKibben recognizes the impulse: "Objecting even slightly to immortality," he rightly says, "is a little like arguing against ice cream—eternal life has only been humanity's great dream since the moment we became conscious." But unlike all previous generations, ours might be able to achieve it. This would alter us beyond recognition. We'd become a different species—one living in eternal bliss, in the eyes of its proponents; sort of like—well, angels, or superhuman beings, anyway. It would certainly mean an end to narrative. If life is endless, why tell stories? No more beginnings and middles because there will be no more endings. No Shakespeare for us, or Dante, or, well, any art, really. It's all infested with mortality and reeks of earthiness. Our new angel-selves will no longer need or understand our art. They might have other art, though it would be pretty bloodless.

But once we're well and truly immortal, what would we do all day? Wouldn't we get tired of the endlessness, the monotony, the lack of meaningful event? Wouldn't we get bored? Nope. We'd sit around and contemplate problems such as: "Where did the universe come from?" "Why is there something rather than nothing?" "What is the meaning of conscious existence?" Is that to be the result of all this admittedly fascinating science—a tedious first-year philosophy seminar? "Not to be impolite," says McKibben, "but for this we trade our humanity?"

That's the good version of the immortal mind. I encountered the bad version in a paperback I received through a high school book-a-month club. *Donovan's Brain* was its title, and the brain in question was being kept alive in a large fish tank and fed on brain food. The hope of the scientists doing this was that the brain would grow in power and strength, and solve problems, such as "Why is there something rather than nothing?" and benefit humankind. But Donovan when he had a body was a stock manipulator or the equivalent, and he bent his newfound mind powers in the direction of world domination, zapping people who got in his way. A big brain does not mean a nice brain. This was made clear to me at the age of twelve, and it's made even clearer in *Enough.* There are some very clever people at work on the parts that will go into making up our immortality, and what they're doing is on some levels fascinating—like playing with the biggest toy box you've ever seen—but they are not the people who should be deciding our future. Asking these kinds of scientists what improved human nature should be like is like asking ants what you should have in your backyard. Of course they would say "more ants."

And while we're on the subject, who exactly is "we"? The "we," that is, who are promised all these goodies. "We" will be the "GenRich," the rich in genes. "We" are certainly not the six billion people already on the planet, nor the ten billion projected for the year 2050—those will be the "GenPoor." "We," when we appear, will be a select few, and since our enhanced genes and our immortality are going to be so expensive, and will not survive—for instance—being squashed flat by tanks, we will have to take steps to protect ourselves. Doubtless "we" will devise almost impenetrable walls, as in the Zamyatin novel of the same name, or "we" will live in a castle, with "them"—the serfs and peasants, the dimwits, the mortals—roiling around outside. We will talk like James Dewey Watson; we'll say things like "It's not much fun being around dumb people." In fact, we'll behave a lot like the aristocrats of old, convinced of our own divine right. The serfs and peasants will hate us. Not to throw cold water on it, but if the serfs and peasants are true to form, sooner or later they'll get hold of some pitchforks and torches and storm the barricades. So to avoid the peasants, we'll have to go into outer space. Having fun yet?

The agenda of those who visualize themselves as the GenRich—like Past Lifers, Future Lifers never see themselves playing the role of ditch digger—is being pushed in the name of that magic duo, progress and inevitability, the twins that always make an appearance when quite a few potential shareholders smell megabucks in the air. (Along with them come the usual my-dick-is-bigger adjectives, as McKibben points out—guts and risk-taking and so forth—so if you don't rush out and get your genes spliced and your head frozen, you're some sort of a wuss.) "Progress" has deluded many, but surely its pretensions as a rallying slogan have been exploded by now. As for "inevitability," it's the rapist's argument: the thing is going to happen anyway, so why not just lie back and enjoy it? Resistance is futile. (That was the old advice: now you're told to scream and vomit, thus influencing the outcome. Times change.)

McKibben takes on both of the magic twins, and is particularly moving on "inevitability." We still have choice, he says. Just because a thing has been invented doesn't mean you have to use it. He offers as exempla the atomic bomb; the Japanese samurais' rejection of guns; the Chinese abandonment of advanced sea power; and the Amish, who examine each new technology and accept or reject it according to social and spiritual criteria. We, too, he says, can accept or reject according to social and spiritual criteria. We can, and we should. We must decide as ourselves—as who we already are as human beings. We must decide from the fullness of our present humanity, flawed though it may be. As I've said, McKibben is an optimist. I agree with him about what we should do, but I'm not too sure we'll do it.

The fact is—and this is not an argument McKibben uses explicitly—that the argument for the perfectibility of humankind rests on a logical fallacy. Thus: man is by definition imperfect, say those who would perfect him. But those who would perfect him are themselves, by their own definition, imperfect. And imperfect beings cannot make perfect decisions. The decision about what constitutes human perfection would have to be a perfect decision; otherwise the result would be not perfection but imperfection. As witness the desire for several different mouths.

Perhaps our striving for perfection should take a different, more Blakean form. Perhaps Infinity can be seen in a grain of sand, and Eternity in an hour. Perhaps happiness is not a goal but a road. Perhaps the pursuit of happiness is that happiness. Perhaps we should take a clue from Tennyson, and separate wisdom and knowledge, and admit that wisdom cannot be cloned or manufactured. Perhaps that admission is wisdom. Perhaps enough should be enough for us. Perhaps we should leave well enough alone.

George Orwell: Some Personal Connections

I grew up with George Orwell. I was born in 1939, and *Animal Farm* was published in 1945. Thus I was able to read it at age nine. It was lying around the house, and I mistook it for a book about talking animals, sort of like *The Wind in the Willows.* I knew nothing about the kind of politics in the book—the child's version of politics then, just after the war, consisted of the simple notion that Hitler was bad but dead. So I gobbled up the adventures of Napoleon and Snowball, the smart, greedy, upwardly mobile pigs, and Squealer the spin-doctor, and Boxer the noble but thick-witted horse, and the easily led, slogan-chanting sheep, without making any connection with historical events.

To say that I was horrified by this book would be an understatement. The fate of the farm animals was so grim, the pigs were so mean and mendacious and treacherous, the sheep were so stupid. Children have a keen sense of injustice, and this was the thing that upset me the most: the pigs were so *unjust.* I cried my eyes out when Boxer the horse had an accident and was carted off to be made into dog food instead of being given the quiet corner of the pasture he'd been promised.

The whole experience was deeply disturbing to me, but I am forever grateful to George Orwell for alerting me early to the danger flags I've tried to watch out for since. In the world of *Animal Farm*, most speechifying and public palaver is bullshit and instigated lying, and though many characters are good-hearted and mean well, they can be frightened into closing their eyes to what's really going on. The pigs browbeat the others with ideology, then twist that ideology to suit their own purposes: their language games were evident to me even at that age. As Orwell taught, it isn't the labels—Christianity, socialism, Islam, democracy, Two Legs Bad, Four Legs Good, the works—that are definitive, but the acts done in their names.

I could see, too, how easily those who have toppled an oppressive power take on its trappings and habits. Jean-Jacques Rousseau was right to warn us that democracy is the hardest form of government to maintain; Orwell knew that in the marrow of his bones because he'd seen it in action. How quickly the precept "All Animals Are Equal" is changed into "All Animals Are Equal, but Some Are More Equal Than Others." What oily concern the pigs show for the welfare of the other animals, a concern that disguises their contempt for those they are manipulating. With what alacrity do they put on the once-despised uniforms of the tyrannous humans they have overthrown, and learn to use their whips. How self-righteously they justify their actions, helped by the verbal web-spinning of Squealer, their nimble-tongued press agent, until all power is in their trotters, and pretence is no longer necessary, and they rule by naked force. A *revolution* often means only that: a revolving, a turn of the wheel of fortune, by which those who were at the bottom mount to the top and assume the choice positions, crushing the former power-holders beneath them. We should beware of all those who plaster the landscape with large portraits of themselves, like the evil pig Napoleon.

Animal Farm is one of the most spectacular Emperor-Has-No-Clothes books of the twentieth century, and it got George Orwell into trouble accordingly. People who run counter to the current popular wisdom, who point out the uncomfortably obvious, are likely to be strenuously baa-ed at by herds of angry sheep. I didn't have all that figured out at the age of nine, of course—not in any conscious way. But we learn the patterns of stories before we learn their meanings, and *Animal Farm* has a very clear pattern.

Then along came *Nineteen Eighty-Four,* which was published in 1949. Thus I read it in paperback a couple of years later, when I was in high school. Then I read it again, and again: it was right up there among my favourite books, along with *Wuthering Heights.* At the same time, I absorbed its two companions, Arthur Koestler's *Darkness at Noon* and Aldous Huxley's *Brave New World.* I was keen on all three of them, but I understood *Darkness at Noon* to be a tragedy about events that had already happened, and *Brave New World* to be a satirical comedy, with events that were unlikely to unfold in exactly that way. ("Orgy-Porgy," indeed.) But *Nineteen Eighty-Four* struck me as more realistic, probably because Winston Smith was more like me, a skinny person who got tired a lot and was subjected to physical education under chilly conditions—this was a feature of my school—and who was silently at odds with the ideas and the manner of life proposed for him. (This may be one of the reasons *Nineteen Eighty-Four* is best read when you are an adolescent; most adolescents feel like that.) I sympathized particularly with Winston Smith's desire to write his forbidden thoughts down in a deliciously tempting secret blank book: I myself had not yet started to write, but I could see the attractions of it. I could also see the dangers because it's this scribbling of his—along with illicit sex, another item with considerable allure for a teenager of the 1950s—that gets Winston into such a mess.

Animal Farm charts the progress of an idealistic movement of liberation toward a totalitarian dictatorship headed by a despotic tyrant; *Nineteen Eighty-Four* describes what it's like to live entirely within such a system. Its hero, Winston Smith, has only fragmentary memories of what life was like before the present dreadful regime set in: he's an orphan, a child of the collectivity. His father died in the war that has ushered in the repression, and his mother has disappeared, leaving him with only the reproachful glance she gave him as he betrayed her over a chocolate bar—a small betrayal that acts both as the key to Winston's character and as a precursor to the many other betrayals in the book.

The government of Airstrip One, Winston's "country," is brutal. The constant surveillance, the impossibility of speaking frankly to anyone, the looming, ominous figure of Big Brother, the regime's need for enemies and wars—fictitious though both may be—which are used to terrify the people and unite them in hatred, the mind-numbing slogans, the distortions of language, the destruction of what has really happened by stuffing any record of it down the Memory Hole—these made a deep impression on me. Let me restate that: they frightened the stuffing out of me. Orwell was writing a satire about Stalin's Soviet Union, a place about which I knew very little at the age of fourteen, but he did it so well that I could imagine such things happening anywhere.

There is no love interest in *Animal Farm*, but there is one in *Nineteen Eighty-Four.* Winston finds a soulmate in Julia, outwardly a devoted Party fanatic, secretly a girl who enjoys sex and makeup and other spots of decadence. But the two lovers are discovered, and Winston is tortured for thoughtcrime: inner disloyalty to the regime. He feels that if he can only remain faithful in his heart to Julia, his soul will be saved—a romantic concept, though one we are likely to endorse. But like all absolutist governments and religions, the Party demands that every personal loyalty be sacrificed to it and replaced with an absolute loyalty to Big Brother. Confronted with his worst fear in the dreaded Room 101, where there's a nasty device involving a cage full of starving rats that can be fitted to the eyes, Winston breaks—"Don't do it to me," he pleads, "do it to Julia." (This sentence has become shorthand in our household for the avoidance of onerous duties. Poor Julia—how hard we would make her life if she actually existed. She'd have to be on a lot of panel discussions, for instance.)

After his betrayal of Julia, Winston Smith becomes a handful of malleable goo. He truly believes that two and two make five and that he loves Big Brother. Our last glimpse of him shows him sitting drink-sodden at an outdoor café, knowing he's a dead man walking and having learned that Julia has betrayed him too, while he listens to a popular refrain: "Under the spreading chestnut tree/ I sold you and you sold me."

Orwell has been accused of bitterness and pessimism—of leaving us with a vision of the future in which the individual has no chance, and the brutal, totalitarian boot of the all-controlling Party will grind into the human face forever. But this view of Orwell is contradicted by the last chapter in the book, an essay on Newspeak—the doublethink language concocted by the regime. By expurgating all words that might be troublesome—"bad" is no longer permitted but becomes "double-plus-ungood"—and by making other words mean the opposite of what they used to mean—the place where people get tortured is the Ministry of Love, the building where the past is destroyed is the Ministry of Information—the rulers of Airstrip One wish to make it literally impossible for people to think straight. However, the essay on Newspeak is written in standard English, in the third person, and in the past tense, which can only mean that the regime has fallen, and that language and individuality have survived. For whoever has written the essay on Newspeak, the world of *Nineteen Eighty-Four* is over. Thus it's my view that Orwell had much more faith in the resilience of the human spirit than he's usually been given credit for.

Orwell became a direct model for me much later in my life—in the real 1984, the year in which I began writing a somewhat different dystopia, *The Handmaid's Tale*. By that time I was forty-four, and I'd learned enough about real despotisms—through the reading of history, through travel, and through my membership in Amnesty International—that I didn't need to rely on Orwell alone.

The majority of dystopias—Orwell's included—have been written by men, and the point of view has been male. When women have appeared in them, they have been either sexless automatons or rebels who've defied the sex rules of the regime. They've acted as the temptresses of the male protagonists, however welcome this temptation may be to the men themselves. Thus Julia, thus the cami-knicker-wearing, orgy-porgy seducer of the Savage in *Brave New World*, thus the subversive femme fatale of Yevgeny Zamyatin's 1924 seminal classic, *We*. I wanted to try a dystopia from the female point of view—the world according to Julia, as it were. However, this does not make *The Handmaid's Tale* a "feminist dystopia," except insofar as giving a woman a voice and an inner life will always be considered "feminist" by those who think women ought not to have these things.

In other respects, the despotism I describe is the same as all real ones and most imagined ones. It has a small, powerful group at the top that controls—or tries to control—everyone else, and it gets the lion's share of available goodies. The pigs in *Animal Farm* get the milk and the apples, the elite of *The Handmaid's Tale* get the fertile women. The force that opposes the tyranny in my book is one in which Orwell himself—despite his belief in the need for political organization to combat oppression—always put great store: ordinary human decency, of the kind he praised in his essay on Charles Dickens. The biblical expression of this quality is probably in the verse "Insofar as you do it unto the least of these, you do it unto me." Tyrants and the powerful believe, with Lenin, that you can't make an omelette without breaking eggs and that the end justifies the means. Orwell, when push came to shove, would have believed—on the contrary—that the means define the end. He wrote as if he sided with John Donne, who said, "Every man's death diminishes me." And so say—I would hope—all of us.

At the end of *The Handmaid's Tale*, there's a section that owes much to *Nineteen Eighty-Four.* It's the account of a symposium held several hundred years in the future, in which the repressive government described in the novel is now merely a subject for academic analysis. The parallels with Orwell's essay on Newspeak should be evident.

Orwell has been an inspiration to generations of writers in another important respect—his insistence on the clear and exact use of language. "Prose like a windowpane," he said, opting for plainsong rather than ornament. Euphemisms and skewed terminology should not obscure the truth. "Acceptable megadeaths" rather than "millions of rotting corpses," but hey, it's not us who're dead; "untidiness" instead of "massive destruction"—this is the beginning of Newspeak. Fancy verbiage is what confuses Boxer the horse and underpins the chantings of the sheep. To insist on *what is*, in the face of ideological spin, popular consensus, and official denial: Orwell knew this takes honesty, and a lot of guts. The position of odd man out is always an uneasy one, but the moment we look around and find that there are no longer any odd men among our public voices is the moment of most danger—because that's when we'll be in lockstep, ready for the Three Minutes' Hate.

The twentieth century could be seen as a race between two versions of man-made Hell—the jackbooted state totalitarianism of Orwell's *Nineteen Eighty-Four* and the hedonistic ersatz paradise of *Brave New World*, where absolutely everything is a consumer good and human beings are engineered to be happy. With the fall of the Berlin Wall in 1989, it seemed for a time that *Brave New World* had won—from henceforth, state control would be minimal, and all we'd have to do was go shopping and smile a lot, and wallow in pleasures, popping a pill or two when depression set in.

But with the notorious 9/11 World Trade Center and Pentagon attacks in 2001, all that changed. Now it appears we face the prospect of two contradictory dystopias at once—open markets, closed minds—because state surveillance is back again with a vengeance. The torturer's dreaded Room 101 has been with us for millennia. The dungeons of Rome, the Inquisition, the Star Chamber, the Bastille, the proceedings of General Pinochet and of the junta in Argentina—all have depended on secrecy and on the abuse of power. Lots of countries have had their versions of it—their ways of silencing troublesome dissent. Democracies have traditionally defined themselves by, among other things, openness and the rule of law. But now it seems that we in the West are tacitly legitimizing the methods of the darker human past, upgraded technologically and sanctified to our own uses, of course. For the sake of freedom, freedom must be renounced. To move us toward the improved world—the utopia we're promised—dystopia must first hold sway. It's a concept worthy of doublethink. It's also, in its ordering of events, strangely Marxist. First the dictatorship of the proletariat, in which lots of heads must roll; then the pie-in-the-sky classless society, which oddly enough never materializes. Instead we just get pigs with whips.

What would George Orwell have to say about it? I often ask myself.

Quite a lot.

Ten Ways of Looking at *The Island of Doctor Moreau* by H. G. Wells

H. G. Wells's *The Island of Doctor Moreau* is one of those books that, once read, is rarely forgotten. Jorge Luis Borges called it an "atrocious miracle" and made large claims for it. Speaking of Wells's early tales—*The Island of Doctor Moreau* among them—he said, "I think they will be incorporated, like the fables of Theseus or Ahasuerus, into the general memory of the species and even transcend the fame of their creator or the extinction of the language in which they were written."

This has proved true, if film may be considered a language unto itself. *The Island of Doctor Moreau* has inspired three films—two of them quite bad—and doubtless few who saw them remembered that it was Wells who authored the book. The story has taken on a life of its own, and, like the offspring of Mary Shelley's *Frankenstein*, has acquired attributes and meanings not present in the original. Moreau himself, in his filmic incarnations, has drifted toward the type of the Mad Scientist, or the Peculiar Genetic Engineer, or the Tyrant-in-Training, bent on taking over the world; whereas Wells's Moreau is certainly not mad, is a mere vivisectionist, and has no ambitions to take over anything whatsoever.

Borges's use of the word *fable* is suggestive, for—despite the realistically rendered details of its surface—the book is certainly not a novel, if by that we mean a prose narrative dealing with observable social life. "Fable" points to a certain folkloric quality that lurks in the pattern of this curious work, as animal faces may lurk in the fronds and flowers of an Aubrey Beardsley design. The term may also indicate a lie—something fabulous or invented, as opposed to that which demonstrably exists—and employed this way it is quite apt, as no man ever did or ever will turn animals into human beings by cutting them up and sewing them together again. In its commonest sense, a fable is a tale—like those of Aesop—meant to convey some useful lesson. But what is that useful lesson? It is certainly not spelled out by Wells.

"Work that endures is always capable of an infinite and plastic ambiguity; it is all things for all men," says Borges, ". . . and it must be ambiguous in an evanescent and modest way, almost in spite of the author; he must appear to be ignorant of all symbolism. Wells displayed that lucid innocence in his first fantastic exercises, which are to me the most admirable part of his admirable work." Borges carefully did not say that Wells employed no symbolism: only that he appeared to be ignorant of doing so.

Here follows what I hope will be an equally modest attempt to probe beneath the appearance, to examine the infinite and plastic ambiguity, to touch on the symbolism that Wells may or may not have employed deliberately, and to try to discover what the useful lesson—if there is one—might be.

TEN WAYS OF LOOKING AT *THE ISLAND OF DOCTOR MOREAU*

1. ELOIS AND MORLOCKS

The Island of Doctor Moreau was published in 1896, when H. G. Wells was only thirty years old. It followed *The Time Machine*, which had appeared the year before, and was to be followed two years later by *The War of the Worlds*, this being the book that established Wells as a force to be reckoned with at a mere thirty-two years of age.

To some of literature's more gentlemanly practitioners—those, for instance, who had inherited money and didn't have to make it by scribbling—Wells must have seemed like a puffed-up little counter-jumper, and a challenging one at that, because he was bright. He'd come up the hard way. In the stratified English social world of the time, he was neither working class nor top crust. His father was an unsuccessful tradesman; he himself apprenticed with a draper for two years before wending his way, via school-teaching and a scholarship, to the Normal School of Science. Here he studied under Darwin's famous apologist, Thomas Henry Huxley. He graduated with a first-class degree, but he'd been seriously injured by one of the students while teaching, an event that put him off schoolmastering. It was after this that he turned to writing.

The Time Traveller in *The Time Machine*—written just before *The Island of Doctor Moreau*—finds that human beings in the future have split into two distinct races. The Eloi are as pretty as butterflies but useless; the grim and ugly Morlocks live underground, make everything, and come out at night to devour the Eloi, whose needs they also supply. The upper classes, in other words, have become a bevy of upper-class twitterers and have lost the ability to fend for themselves, and the working classes have become vicious and cannibalistic.

Wells was neither an Eloi nor a Morlock. He must have felt he represented a third way, a rational being who had climbed up the ladder through ability alone, without partaking of the foolishness and impracticality of the social strata above his nor of the brutish crudeness of those below.

But what about Prendick, the narrator of *The Island of Doctor Moreau*? He's been pootling idly about the world, for his own diversion we assume, when he's shipwrecked. The ship is called the *Lady Vain*, surely a comment on the snooty aristocracy. Prendick himself is a "private gentleman" who doesn't have to work for a living, and though he—like Wells—has studied with Huxley, he has done so not out of necessity but out of dilettantish boredom—"as a relief from the dullness of [his] comfortable independence." Prendick, though not quite as helpless as a full-fledged Eloi, is well on the path to becoming one. Thus his hysteria, his lassitude, his moping, his ineffectual attempts at fair play, and his lack of common sense—he can't figure out how to make a raft because he's never done "any carpentry or suchlike work" in his life, and when he does manage to patch something together, he's situated it too far from the sea and it falls apart when he's dragging it. Although Prendick is not a complete waste of time—if he were, he wouldn't be able to hold our attention while he tells his story—he's nonetheless in the same general league as the weak-chinned curate in the later *The War of the Worlds*, that helpless and drivelling "spoiled child of life."

His name—Prendick—is suggestive of "thick" coupled with

"prig," this last a thing he is explicitly called. To those versed in legal lore, it could suggest *prender*, a term for something you are empowered to take without it having been offered. But it more nearly suggests *prentice*, a word that would have been floating close to the top of Wells's semiconsciousness, due to his own stint as an apprentice. Now it's the upper class's turn at apprenticeship! Time for one of them to undergo a little degradation and learn a thing or two. But what?

2. SIGNS OF THE TIMES

The Island of Doctor Moreau not only comes midway in Wells's most fertile period of fantastic inventiveness, it also comes during such a period in English literary history. Adventure romance had taken off with Robert Louis Stevenson's *Treasure Island* in 1882, and H. Rider Haggard had done him one better with *She* in 1887. This latter coupled straight adventure—shipwreck, tramps through dangerous swamps and nasty shrubbery, encounters with bloody-minded savages, fun in steep ravines and dim grottos—with a big dollop of weirdness carried over from earlier Gothic traditions, done up this time in a package labelled "Not Supernatural." The excessive powers of She are ascribed not to a close encounter with a vampire or god but to a dip in a revolving pillar of fire, no more supernatural than lightning. She gets her powers from Nature.

It's from this blend—the grotesque and the "natural"—that Wells took his cue. An adventure story that would once have featured battles with fantastic monsters—dragons, gorgons, hydras—keeps the exotic scenery, but the monsters have been produced by the very agency that was seen by many in late

Victorian England as the bright, new, shiny salvation of humankind: science.

The other blend that proved so irresistible to readers was one that was developed much earlier, and to singular advantage, by Jonathan Swift: a plain, forthright style in the service of incredible events. Poe, that master of the uncanny, piles on the adjectives to create "atmosphere"; Wells, on the other hand, follows R. L. Stevenson and anticipates Hemingway in his terse, almost journalistic approach, usually the hallmark of the ultra-realists. *The War of the Worlds* shows Wells employing this combination to best effect—we think we're reading a series of news reports and eyewitness accounts—but he's already honing it in *The Island of Doctor Moreau.* A tale told so matter-of-factly and with such an eye to solid detail surely cannot be—we feel—either an invention or a hallucination.

3. SCIENTIFIC

Wells is acknowledged to be one of the foremost inventors in the genre we now know as "science fiction." As Robert Silverberg has said, "Every time-travel tale written since *The Time Machine* is fundamentally indebted to Wells. . . . In this theme, as in most of science fiction's great themes, Wells was there first."

Science fiction as a term was unknown to Wells; it did not make its appearance until the 1930s, in America, during the golden age of bug-eyed monsters and girls in brass brassieres. Wells himself referred to his science-oriented fictions as "scientific romances"—a term that did not originate with him but with a lesser-known writer called Charles Howard Hinton.

There are several interpretations of the term *science*. If it implies the known and the possible, then Wells's scientific romances are by no means scientific; he paid little attention to those boundaries. The "science" part of these tales is embedded instead in a worldview that derived from Wells's study of Darwinian principles under Huxley and has to do with the grand study that engrossed him throughout his career: the nature of man. This, too, may account for his veering between extreme utopianism (if man is the result of evolution, not of Divine creation, surely he can evolve yet further?) and the deepest pessimism (if man came from the animals and is akin to them, rather than to the angels, surely he might slide back the way he came?). *The Island of Doctor Moreau* belongs to the debit side of the Wellsian account book.

Darwin's *On the Origin of Species* and *The Descent of Man* were a profound shock to the Victorian system. Gone was the God who spoke the world into being in seven days and made man out of clay; in his place stood millions of years of evolutionary change and a family tree that included primates. Gone, too, was the kindly Wordsworthian version of Mother Nature that had presided over the first years of the century; in her place was Tennyson's "Nature, red in tooth and claw/ With ravine." The devouring femme fatale that became so iconic in the 1880s and 1890s owes a lot to Darwin. So does the imagery and cosmogony of *The Island of Doctor Moreau*.

4. ROMANCE

So much for the "scientific" in "scientific romance." What about the "romance"?

In both "scientific romance" and "science fiction," the scientific element is merely an adjective; the nouns are "romance" and "fiction." In respect to Wells, "romance" is more helpful than "fiction."

"Romance," in today's general usage, is what happens on Valentine's Day. As a literary term it has slipped in rank somewhat—being now applied to such things as Harlequin Romances—but it was otherwise understood in the nineteenth century when it was used in opposition to the term *novel.* The novel dealt with known social life, but a romance could deal with the long ago and the far away. It also allowed much more latitude in terms of plot. In a romance, event follows exciting event at breakneck pace. As a rule, this has caused the romance to be viewed by the high literati—those bent more on instruction than on delight—as escapist and vulgar, a judgment that goes back at least two thousand years.

In *The Secular Scripture,* Northrop Frye provides an exhaustive analysis of the structure and elements of the romance as a form. Typically a romance begins with a break in ordinary consciousness, often—traditionally—signalled by a shipwreck, frequently linked with a kidnapping by pirates. Exotic climes are a feature, especially exotic desert islands; so are strange creatures.

In the sinister portions of a romance, the protagonist is often imprisoned or trapped, or lost in a labyrinth or maze, or in a forest that serves the same purpose. Boundaries between the normal levels of life dissolve: vegetable becomes animal, animal becomes quasi-human, human descends to animal. If the lead character is female, an attempt will be made on her virtue, which she manages miraculously to preserve. A rescue, however improbable, restores the protagonist to his or her previous life and reunites him or her with loved ones. *Pericles, Prince of Tyre* is a romance. It's got everything but talking dogs.

The Island of Doctor Moreau is also a romance, though a dark one. Consider the shipwreck. Consider the break in the protagonist's consciousness—the multiple breaks, in fact. Consider the pirates, here supplied by the vile captain and crew of the *Ipecacuanha.* Consider the name *Ipecacuanha*, signifying an emetic and purgative: the break in consciousness is going to have a nastily physical side to it, of a possibly medicinal kind. Consider the fluid boundaries between animal and human. Consider the island.

5. THE ENCHANTED ISLAND

The name given to the island by Wells is Noble's Island, a patent irony as well as another poke at the class system. Say it quickly and slur a little, and it's *no blessed island.*

This island has many literary antecedents and several descendants. Foremost among the latter is William Golding's island in *Lord of the Flies*—a book that owes something to *The Island of Doctor Moreau*, as well as to those adventure books *Coral Island* and *The Swiss Family Robinson*, and of course to the great original shipwreck-on-an-island classic, *Robinson Crusoe. Moreau* could be thought of as one in a long line of island-castaway books.

All those just mentioned, however, keep within the boundaries set by the possible. *The Island of Doctor Moreau* is, on the contrary, a work of fantasy, and its more immediate grandparents are to be found elsewhere. *The Tempest* springs immediately to mind: here is a beautiful island, belonging at first to a witch, then taken over by a magician who lays down the law, particularly to the malignant, animal-like Caliban, who will obey only when pain is inflicted on him. Dr. Moreau could be seen as a sinister version of Prospero, surrounded by a hundred or so Calibans of his own creation.

But Wells himself points us toward another enchanted island. When Prendick mistakenly believes that the beast-men he's seen were once men, he says, "[Moreau] had merely intended . . . to fall upon me with a fate more horrible than death, with torture, and after torture the most hideous degradation it was possible to conceive—to send me off, a lost soul, a beast, to the rest of [the] Comus rout."

Comus, in the masque of that name by Milton, is a powerful sorcerer who rules a labyrinthine forest. He's the son of the enchantress Circe, who in Greek myth was the daughter of the Sun and lived on the island of Aeaea. Odysseus landed there during his wanderings, and Circe transformed his crew into pigs. She has a whole menagerie of other kinds of animals—wolves, lions—that were also once men. Her island is an island of transformation: man to beast (and then to man again, once Odysseus gets the upper hand).

As for Comus, he leads a band of creatures, once men, who have drunk from his enchanted cup and have turned into hybrid monsters. They retain their human bodies, but their heads are those of beasts of all kinds. Thus changed, they indulge in sensual revels. Christina Rossetti's *Goblin Market*, with its animal-form goblins who tempt chastity and use luscious edibles as bait, is surely a late offshoot of *Comus*.

As befits an enchanted island, Moreau's island is both semi-alive and female but not in a pleasant way. It's volcanic, and emits from time to time a sulphurous reek. It comes equipped with flowers, and also with clefts and ravines, fronded on either side. Moreau's beast-men live in one of these, and since they do not have very good table manners, it has rotting food in it and it smells bad. When the beast-men start to lose their humanity and revert to their beast natures, this locale becomes the site of a moral breakdown that is specifically sexual.

What is it that leads us to believe that Prendick will never have a girlfriend?

6. THE UNHOLY TRINITY

Nor will Dr. Moreau. There is no Mrs. Moreau on the island. There are no female human beings at all.

Similarly, the God of the Old Testament has no wife. Wells called *The Island of Doctor Moreau* "a youthful piece of blasphemy," and it's obvious that he intended Moreau—that strong, solitary gentleman with the white hair and beard—to resemble traditional paintings of God. He surrounds Moreau with semi-biblical language as well: Moreau is the lawgiver of the island; those of his creatures who go against his will are punished and tortured; he is a god of whim and pain. But he isn't a real God because he cannot really create; he can only imitate, and his imitations are poor.

What drives him on? His sin is the sin of pride, combined with a cold "intellectual passion." He wants to know everything. He wishes to discover the secrets of life. His ambition is to be as God the Creator. As such, he follows in the wake of several other aspirants, including Dr. Frankenstein and Hawthorne's various alchemists. Dr. Faustus hovers in the background, but he wanted youth and wealth and sex in return for his soul, and Moreau has no interest in such things; he despises what he calls "materialism," which includes pleasure and pain. He dabbles in bodies but wishes to detach himself from his own. (He has some literary brothers: Sherlock Holmes would understand his bloodless intellectual passion. So would Oscar Wilde's Lord Henry Wotton, of that earlier fin-de-siècle transformation novel, *The Picture of Dorian Gray*.)

But in Christianity, God is a trinity, and on Moreau's island there are three beings whose names begin with *M. Moreau* as a name combines the syllable "mor"—from *mors, mortis*, no doubt—with the French for "water," suitable in one who aims at exploring the limits of plasticity. The whole word means "moor" in French. So the very white Moreau is also the Black Man of witchcraft tales, a sort of anti-God.

Montgomery, his alcoholic assistant, has the face of a sheep. He acts as the intercessor between the beast folk and Moreau, and in this function stands in for Christ the Son. He's first seen offering Prendick a red drink that tastes like blood, and some boiled mutton. Is there a hint of an ironic Communion service here—blood drink, flesh of the Lamb? The communion Prendick enters into by drinking the red drink and eating the mutton is the communion of carnivores, that human communion forbidden to the beast folk. But it's a communion he was part of anyway.

The third person of the Trinity is the Holy Spirit, usually portrayed as a dove—God in living but nonhuman form. The third M creature on the island is M'Ling, the beast creature who serves as Montgomery's attendant. He, too, enters into the communion of blood: he licks his fingers while preparing a rabbit for the human beings to eat. The Holy Spirit as a deformed and idiotic man-animal? As a piece of youthful blasphemy, *The Island of Doctor Moreau* was even more blasphemous than most commentators have realized.

Just so we don't miss it, Wells puts a serpent beast into his dubious garden: a creature that was completely evil and very strong, and that bent a gun barrel into the letter *S*. Can Satan, too, be created by man? If so, blasphemous indeed.

7. THE NEW WOMAN AS CATWOMAN

There are no female human beings on Moreau's island, but Moreau is busily making one. The experiment on which he's engaged for most of the book concerns his attempt to turn a female puma into the semblance of a woman.

Wells was more than interested in members of the cat family, as Brian Aldiss has pointed out. During his affair with Rebecca West, she was "Panther," he was "Jaguar." But "cat" has another connotation: in slang, it meant "prostitute." This is Montgomery's allusion when he says—while the puma is yelling under the knife—"I'm damned . . . if this place is not as bad as Gower Street—with its cats." Prendick himself makes the connection explicit on his return to London when he shies away from the "prowling women [who] would mew after me."

"I have some hope of her head and brain," says Moreau of the puma. ". . . I will make a rational creature of my own." But the puma resists. She's almost a woman—she weeps like one—but when Moreau begins torturing her again, she utters a "shriek almost like that of an angry virago." Then she tears her fetter out of the wall and runs away, a great, bleeding, scarred, suffering, female monster. It is she who kills Moreau.

Like many men of his time, Wells was obsessed with the New Woman. On the surface of it he was all in favour of sexual emancipation, including free love, but the freeing of Woman evidently had its frightening aspects. Rider Haggard's *She* can be seen as a reaction to the feminist movement of his day—if women are granted power, men are doomed—and so can Wells's deformed puma. Once the powerful, monstrous sexual cat tears her fetter out of the wall and gets loose, minus the improved brain she ought to have, courtesy of Man the Scientist, look out.

8. THE WHITENESS OF MOREAU, THE BLACKNESS OF M'LING

Wells was not the only nineteenth-century English writer who used furry creatures to act out English sociodramas. Lewis Carroll had done it in a whimsical way in the *Alice* books, Kipling in a more militaristic fashion in *The Jungle Book.*

Kipling made the Law sound kind of noble in *The Jungle Book.* Not so Wells. The Law mumbled by the animal-men in Moreau is a horrible parody of Christian and Jewish liturgy; it vanishes completely when the language of the beasts dissolves, indicating that it was a product of language, not some eternal, extra-lingual, God-given creed.

Wells was writing at a time when the British Empire still held sway, but the cracks were already beginning to show. Moreau's island is a little colonial enclave of the most hellish sort. It's no accident that most (although not all) of the beast folk are black or brown, that they are at first thought by Prendick to be "savages" or "natives," and that they speak in a kind of mangled English. They are employed as servants and slaves, in a regime that's kept in place with whip and gun; they secretly hate the real "men" as much as they fear them; and they disobey the Law as much as possible, and kick over the traces as soon as they can. They kill Moreau and they kill Montgomery and they kill M'Ling, and, unless Prendick can get away, they will kill him too, although at first he "goes native" and lives among them, and does things that fill him with disgust and that he would rather not mention.

White Man's Burden, indeed.

9. THE MODERN ANCIENT MARINER

The way in which Prendick escapes from the island is noteworthy. He sees a small boat with a sail and lights a fire to hail it. It approaches, but strangely: it doesn't sail with the wind, but yaws and veers. There are two figures in it, one with red hair. As the boat enters the bay, "Suddenly a great white bird flew up out of the boat, and neither of the men stirred nor noticed it. It circled round, and then came sweeping overhead with its strong wings outspread." This bird cannot be a gull: it's too big and solitary. The only white seabird usually described as "great" is the albatross.

The two figures in the boat are dead. But it is this death boat, this life-in-death coffin boat, that proves the salvation of Prendick.

In what other work of English literature do we find a lone man reduced to a pitiable state, a boat that sails without a wind, two death figures, one with unusual hair, and a great white bird? The work is, of course, *The Ancient Mariner,* which revolves around man's proper relation to Nature and concludes that this proper relation is one of love. It is when he manages to bless the sea serpents that the Mariner is freed from the curse he has brought upon himself by shooting the albatross.

The Island of Doctor Moreau also revolves around man's proper relation to Nature, but its conclusions are quite different because Nature itself is seen differently. It is no longer the benevolent, motherly Nature eulogized by Wordsworth, for between Coleridge and Wells came Darwin.

The lesson learned by the albatross-shooting Mariner is summed up by him at the end of the poem:

> He prayeth well, who loveth well
> Both man and bird and beast.
> He prayeth best, who loveth best
> All things both great and small;
> For the dear God who loveth us,
> He made and loveth all.

In *The Ancient Mariner*–like pattern at the end of *The Island of Doctor Moreau*, the "albatross" is still alive. It has suffered no harm at the hands of Prendick. But he lives in the shadow of a curse anyway. His curse is that he can't love or bless anything living—not bird, not beast, and most certainly not man. He has another curse too: the Ancient Mariner is doomed to tell his tale, and those who are chosen to hear it are convinced by it. But Prendick chooses not to tell because, when he tries, no one will believe him.

What then is the lesson learned by the unfortunate Prendick? It can perhaps best be understood in reference to *The Ancient Mariner.* The god of Moreau's island can scarcely be described as a dear God who makes and loves all creatures. If Moreau is seen to stand for a version of God the Creator who "makes" living things, he has done, in Prendick's final view, a very bad job. Similarly, if God can be considered as a sort of Moreau, and if the equation "Moreau is to his animals as God is to man" may stand, then God himself is accused of cruelty and indifference—making man for fun and to satisfy his own curiosity and pride, laying laws on him he cannot understand or obey, then abandoning him to a life of torment.

Prendick cannot love the distorted and violent furry folk on the island, and it's just as hard for him to love the human beings he encounters on his return to "civilization." Like Swift's Gulliver, he can barely stand the sight of his fellow men. He lives in a state of queasy fear, inspired by his continued experience of dissolving boundaries: as the beasts on the island have at times appeared human, the human beings he encounters in England appear bestial. He displays his modernity by going to a "mental specialist," but this provides only a partial remedy. He feels himself to be "an animal tormented . . . sent to wander alone."

Prendick forsakes his earlier dabblings in biology and turns instead to chemistry and astronomy. He finds "hope"—"a sense of infinite peace and protection"—in "the glittering hosts of heaven." As if to squash even this faint hope, Wells almost immediately wrote *The War of the Worlds,* in which not peace and protection, but malice and destruction, come down from the heavens in the form of the monstrous but superior Martians.

The War of the Worlds can be read as a further gloss on Darwin. Is this where evolution will lead—to the abandonment of the body, to giant, sexless, blood-sucking heads with huge brains and tentaclelike fingers? But it can also be read as a thoroughly chilling coda to *The Island of Doctor Moreau.*

NOTES

150 Jorge Luis Borges, *Other Inquisitions*, p. 87.

153 *The War of the Worlds*, p. 117.

155 Silverberg, *Voyagers in Time*, p. x.

155 The "brass brassiere" is from an oral history of science fiction prepared by Richard Wolinsky for Berkeley's KPFA-FM.

Never Let Me Go
by Kazuo Ishiguro

Never Let Me Go is the sixth novel by Kazuo Ishiguro, who won the Booker Prize in 1989 for his chilling rendition of a bootlickingly devoted but morally blank English butler, *The Remains of the Day*. It's a thoughtful, crafty, and finally very disquieting look at the effects of dehumanization on any group that's subject to it. In Ishiguro's subtle hands, these effects are far from obvious. There's no Them-Bad, Us-Good preaching; rather there's the feeling that as the expectations of such a group are diminished, so is its ability to think outside the box it has been shut up in. The reader reaches the end of the book wondering exactly where the walls of his or her own invisible box begin and end.

Ishiguro likes to experiment with literary hybrids, and to hijack popular forms for his own ends, and to set his novels against tenebrous historical backdrops; thus, *When We Were Orphans* mixes the Boy's Own Adventure with the 1930s detective story while taking a whole new slice out of the Second World War. An Ishiguro novel is never about what it pretends to pretend to be about, and *Never Let Me Go* is true to form. You might think of it as the Enid Blyton schoolgirl story crossed with *Blade Runner*, and perhaps also with John Wyndham's shunned-children classic, *The Chrysalids*: the children in Wyndham's novel, like those in *Never Let Me Go*, give other people the creeps.

The narrator, Kathy H., is looking back on her school days at a superficially idyllic establishment called Hailsham. (As in "sham"; as in Charles Dickens's Miss Havisham, exploiter of uncomprehending children.) At first you think the "H" in "Kathy H." is the initial of a surname, but none of the students at Hailsham has a real surname. Soon you understand that there's something very peculiar about this school. Tommy, for instance, who is the best boy at football, is picked on because he's no good at art: in a conventional school it would be the other way around.

In fact, Hailsham exists to raise cloned children who have been brought into the world for the sole purpose of providing organs to other, "normal" people. They don't have parents. They can't have children. Once they graduate, they will go through a period of being "carers" to others of their kind who are already being deprived of their organs; then they will undergo up to four "donations" themselves, until they "complete." (None of these terms has originated with Ishiguro; he just gives them an extra twist.) The whole enterprise, like most human enterprises of dubious morality, is wrapped in euphemism and shadow: the outer world wants these children to exist because it's greedy for the benefits they can confer, but it doesn't wish to look head-on at what is happening. We assume—though it's never stated—that whatever objections might have been raised to such a scheme have already been overcome. By now the rules are in place and the situation is taken for granted—as slavery was once—by beneficiaries and victims alike.

All this is background. Ishiguro isn't much interested in the practicalities of cloning and organ donation. (Which four organs, you may wonder? A liver, two kidneys, then the heart? But wouldn't you be dead after the second kidney anyway? Or are we throwing in the pancreas?) Nor is this a novel about future horrors: it's set not in a Britain-yet-to-come but in a Britain-off-to-the-side, in which cloning has been introduced before the 1970s. Kathy H. is thirty-one in the late 1990s, which places her childhood and adolescence in the 1970s and early 1980s—close to those of Ishiguro, who was born in 1955 in Nagasaki and moved to England when he was five. (Surely there's a connection: as a child, Ishiguro must have seen many young people dying far too soon, through no fault of their own.) And so the observed detail is realistic—the landscapes, the kind of sports pavilion at Hailsham, the assortment of teachers and "guardians," even the fact that Kathy listens to her music via tape, not CD.

Kathy H. has nothing to say about the unfairness of her fate. Indeed, she considers herself lucky to have grown up in a superior establishment like Hailsham rather than on the standard organ farm. Like most people, she's interested in personal relationships: in her case, the connection between her "best friend," the bossy and manipulative Ruth, and the boy she loves—Tommy, the amiable football-playing bad artist. Ishiguro's tone is perfect: Kathy is intelligent but nothing extraordinary, and she prattles on in the obsessive manner touchy girls have, going back over past conversations and registering every comment and twitch and crush and put-down and cold shoulder and gang-up and spat. It's all hideously familiar and gruesomely compelling to anyone who's ever kept a teenaged diary.

In the course of her story, Kathy H. solves a few of the mysteries that have been bothering her. Why is it so important that these children make art, and why is their art collected and taken away? Why does it matter to anyone that they be educated, if they're only going to die young anyway? Are they human or not? There's a chilling echo of the art-making children in Theresienstadt, and of the Japanese children dying of radiation who nevertheless made paper cranes.

What is art for? the characters ask. They link the question to their own circumstances, but surely they speak for anyone with a connection with the arts: What *is* art for? The notion that it ought to be for *something*, that it must serve some clear social purpose—extolling the gods, cheering people up, illustrating moral lessons—has been around at least since Plato, and was tyrannical in the nineteenth century. It lingers with us still, especially when parents and teachers start squabbling over school curricula. Art does turn out to have a purpose in *Never Let Me Go*, but it isn't quite the purpose the characters have been hoping for.

One motif at the very core of *Never Let Me Go* is the treatment of out-groups, and the way out-groups form in-groups, even among themselves. The marginalized are not exempt from doing their own marginalization: even as they die, Ruth and Tommy and the other donors form a proud, cruel little clique, excluding Kathy H. because, not being a donor yet, she can't really understand.

The book is also about our tendency to cannibalize others to make sure we ourselves get a soft ride. Ursula K. Le Guin has a short story called "The Ones Who Walk Away from Omelas," in which the happiness of the many depends absolutely on the arranged unhappiness of the few, and *Never Let Me Go* could be read as a sister text. The children of Hailsham are human sacrifices, offered up on the altar of improved health for the population at large. With babies already being created with a view to their organs—help for an afflicted sibling, for instance—the dilemma of the Hailsham "students" is bound to become more general. Who owns your body? Who therefore is entitled to offer it up? The reluctance of Kathy H. and her pals to really confront what awaits them—pain, mutilation, death—may account for the curious lack of physicality of Kathy's descriptions of their life. Nobody eats anything much in this book, nobody smells anything. We don't know much about what the main characters look like. Even the sex is oddly bloodless. But landscapes, buildings, and the weather are intensely present. It's as if Kathy has invested a lot of her sense of self in things quite far away from her own body, and thus less likely to be injured.

Finally, the book is also about our wish to do well, to attract approval. The children's poignant desire to be patted on the head—to be a "good carer," keeping those from whom organs are being taken from becoming too distressed; to be a "good donor," someone who makes it through all four "donations"—is heartbreaking. This is what traps them in their cage: none of them thinks about running away or revenging themselves upon the "normal" members of society. Ruth takes refuge in grandiose lies about herself, and in daydreams—maybe she'll be allowed to get an office job. Tommy reacts with occasional rage to the unconscionable things being done to him but then apologizes for his loss of control. In Ishiguro's world, as in our own, most people do what they're told.

Tellingly, two words recur again and again. One, as you might expect, is *normal*. The other is *supposed*, as in the last words of the book: "wherever it was that I was supposed to be going." Who defines "normal"? Who tells us what we are supposed to be doing? These questions always become more pressing in times of stress; unless I'm much mistaken, they'll loom ever larger in the next few years.

Never Let Me Go is unlikely to be everybody's cup of tea. The people in it aren't heroic. The ending is not comforting. Nevertheless, this is a brilliantly executed book by a master craftsman who has chosen a difficult subject: ourselves, seen through a glass, darkly.

After the Last Battle: *Visa for Avalon* by Bryher

The novella-length fiction *Visa for Avalon* by the writer who called herself Bryher was first published in 1965 and was reissued by the Paris Press in 2004, before the U.S. presidential election of that year. Since it is set in the future—a future in which violent mass movements are causing uproar and a controlling government is restricting the freedom of ordinary citizens—it was seen by both its publisher, Jan Freeman, and by its introducer, Susan McCabe, as a book with a lot to say about the squeeze being put on liberal democracy by such draconian measures as the Patriot Act in the United States and by similar tendencies elsewhere.

The Paris Press is "a not-for-profit press publishing work by women that has been neglected or misrepresented by the literary world." For this exemplary aim, Bryher is a strong candidate: few twentieth-century women's lives were more interconnected with their own era, and few others displayed her edgy bravery and intellectual curiosity, but she is little-known today.

Bryher was born in England in 1894, thus living through the First World War as a woman in her twenties, through the intellectually exciting 1920s in her thirties, and through the Second World War in her fifties. Her birth name was Annie Ellerman; she renamed herself after one of the Scilly Isles, a place that—in her inner geography—stood for remoteness, adventure, and freedom: as a child, she longed to run away to sea and become a cabin boy. Her family was well-to-do, which gave her the opportunity to develop her many interests.

At the age of twenty she encountered Ezra Pound and Imagism, and through them, the poet H.D. Bryher and H.D. formed a lifelong friendship and sometime partnership, although they did not always live together. The two of them took up psychoanalysis in the 1920s, and Freud and his teachings remained important to Bryher throughout her life. She was a poet, a supporter of the modernists, and a foster mother of the experimental writers and filmmakers of the 1920s and 1930s. With the rise of fascism she foresaw the coming horror, and when it came she devoted herself to rescuing Jews and intellectuals, using her home base in Switzerland as a transfer point. When Switzerland expelled most foreigners in 1940 she went to England, where she lived through the blitz. After the war she published a series of historical novels that were widely read at the time; but from these, *Visa for Avalon* is a departure.

. . . . .

Bryher was seventy-one when *Visa for Avalon* was published. She had eighteen years yet to live—she died in 1983—and several books left to write; still, anything produced by an author of this age cannot help but be retrospective in mood, and *Visa for Avalon* has something of an autumnal feel to it. As the hand of Death readies itself for the definitive knock on the door, the writer toils even harder: *Wait! Wait! I have just this one very important message I need to get across!* A writer's age at the time of a work's composition is never irrelevant: *The Tempest* is not a young man's play, and *Visa for Avalon* is not a young woman's work.

In her introduction, Susan McCabe links *Visa for Avalon* to the twentieth-century dystopian tradition that includes Orwell's *Nineteen Eighty-Four* and Huxley's *Brave New World*, and such a link is not entirely inappropriate. Yet *Visa for Avalon* is very different from either in tone, and—insofar as one can say anything about a writer's intent—in intent as well. It's an odd duck of a book, and placing it within the Orwell-Huxley frame does it a slight disservice: the reader enters the book expecting the kind of specific and quasi-satirical detail that abounds in these works—the religion of Our Ford, the Ministry of Love, the babies grown in bottles, the use of Newspeak, and so forth—but such sardonic, bat-winged flights of invention are not to be found in its pages.

What then is to be found in them? The word *allegory* has been used about *Visa for Avalon*, but it is not an allegory, since its characters and events cannot be interpreted one-for-one. In Spenser's *Faerie Queene*, Una stands for the True Faith, Duessa for the false one, the Faerie Queene herself for Queen Elizabeth I, and so forth; but there are no such connect-the-dots certainties in *Visa for Avalon*. Part of this book's crepuscular charm is that it eludes definition—its arrangement is fuguelike rather than linear or schematic, and it achieves its effects partly by a seemingly artless syncopation of motifs.

Insofar as *Visa for Avalon* has a central character, that character is Robinson, a man of retirement age whom we first discover waking up by the sea. His name connects him with islands (*Robinson Crusoe* and *The Swiss Family Robinson* must have been known to Bryher, who as a child "devoured" everything in print), and Robinson will soon develop a yen for another island—the island nation of Avalon. Robinson is on vacation in a Cornish-sounding village called Trelawney; he's staying at Rose Cottage, owned by his landlady, the widow Mrs. Lilian Blunt. (Those lilies, those roses: the innocence of the garden in Tennyson's *Maud*, before the catastrophe. Bryher, like most writers of her generation—Pound included—used armfuls of images and much stock vocabulary from the Victorian literature they claimed to be heaving out the window. The hand of Tennyson in particular lay heavily upon them, a point to keep in mind when the significance of "Avalon" is considered. The practical "Blunt" of Lilian's last name stands in opposition to the more romantic elements: this is a landlady who cooks eggs and does other bustling, domestic things, but her Blunt element is—we will discover—somewhat of a disguise.)

After a working life that sounds confining and tedious, Robinson has planned to spend his retirement years in Trelawney. He often goes fishing with a younger man called Alex. But the peaceful scene is disrupted when a "Movement" youth group in green uniforms and caterpillar formation worms its way into the village, when Mrs. Blunt is informed by a bureaucrat that her beloved cottage—her lifelong home, her only possession—is to be demolished to make way for a factory access road, and when it swiftly becomes clear that worse upheavals are in store. The rights of individuals are about to be mashed underfoot in the name of collective progress, as the fate of Rose Cottage exemplifies.

Robinson and Lilian decide to flee, not only from Trelawney but from their no-name England-like country itself. They travel to the also no-name city by means of jammed and filthy trains; they attempt to get a couple of the rare visas for the mysterious island nation called Avalon; they are helped in their efforts by Alex, who works at the Avalon consulate; and finally—joined by a couple of other visa-holders—they make it onto the last tiny plane for Avalon, just before the moblike Movement closes the airport and prepares to "laugh at international law."

Along the way, Robinson and his several fellow travellers provide a running commentary on the trends unfolding. What has caused things to go so thoroughly to the bad? Is it overpopulation? The fact that "people are apathetic until it becomes too late," as Alex says? The repression of this or that urge or passion, bound to burst forth in a thuggish fashion? The unconscious desire of the majority for a return to barbarism? Whatever the reason, no good will come of it. "If an individual's right to a place of his own were not respected," Robinson muses, "it was the first link in a chain that would ultimately lead to the elimination of the unwanted by any group that happened to be in power." Lawson—Avalon's representative at the consulate, and a straight arrow, as his name implies—takes the psychoanalytic view: "Why was there so deep an urge for destruction in people . . . They used so much research to introduce automation into everyday life and so little to find out what really went on in a nation's mind."

But all of this political chat—although true enough, and informed by Bryher's considerable experience with head-in-the-sand denial, and psychoanalysis, and the rampages of iron-fisted and destructive political movements and the stifling of the rule of law—seems almost like flotsam on the surface of a different sea. There's a dreamlike air to the narrative, even apart from the this-can't-be-happening sensations that overtake people when they're caught in sudden violent upheavals. What deeper current was carrying Bryher along as she composed this strange book? She knew the work of Kafka, and his name, too, has been used in connection with *Visa for Avalon*—again with some reason: the facelessness of the malevolent forces, their apparent lack of any definite goal, and the pettiness of the bureaucrats representing them recall, perhaps, *The Trial* and *The Castle.*

But you don't name a book *Visa for Avalon* by accident, especially if you are Bryher, composer of historic novels. "Avalon" in Arthurian legend was the place to which King Arthur was boat-lifted after his last battle. Susan McCabe mentions Geoffrey of Monmouth's version, in which Avalon is a sort of apple-filled Edenic island where Arthur will be healed, and also Malory's *Le Morte d'Arthur,* in which Arthur dies there, surrounded by weeping women. She does not, however, invoke the most likely influence on Bryher, who as a voraciously reading child went through her father's library. That late-Victorian library would have contained the works of Tennyson, including his long narrative poem, *Idylls of the King.* What Bryher's novel means to signify is intimately connected with what Tennyson's *Idylls* meant to signify.

The final section of *Idylls,* "The Passing of Arthur," begins with "that last weird battle in the west," a misty affair filled with confusion and the difficulty of being able to identify the enemy—a confusion and a difficulty mirrored in Bryher's book. Both works have to do with a man's fated journey toward his possible death, the collapse of a civilization and its return to savage and lawless ways, the betrayal of the nobility in man, and—underneath these themes—the sadness of getting older and finding yourself surrounded by young people who don't understand what you've lived through or even what you're talking about. Tennyson sounds this note repeatedly in "The Passing of Arthur," which is told by Sir Bedivere in his old age, when he is living among "new men, strange faces, other minds."

The new men, the strange faces, and the other minds are

already a problem for both Lilian and Robinson, quite apart from any green-uniformed Movement. There's a considerable amount of grumbling about "change," and "development," and "progress," and how things aren't the same as they used to be, and about the rudeness of the young, and also that of waitresses—themes not unheard-of among any group of retirees sharing doughnuts at the corner coffee shop. In addition to this, Robinson is a man who from the outset feels that he's no longer of the present day and finds himself resigned to his own end. "Do you know," he says, toward the beginning, "I wish I could step ashore and die with this moment as my memory of earth."

Section Two begins as Robinson takes his "final walk" beside the waves: "They were full of the terror of death, of the return to the caldron of the sagas, where what was finished was swept away and new patterns formed from the atoms . . . Age was rather an exhaustion of the emotions than a physical fatigue . . . " Or, as Tennyson's Arthur put it, "The old order changeth, yielding place to new . . . " Will Avalon—for Robinson—be a place of death or a place of healing, or will the latter be a version of the former?

And what will Avalon be for the other folk headed there? Each one of the travellers has different hopes connected with it. A young girl thinks happily of love, Alex wants "the truth," the pilot—an Avalonian—is torn between the domestic and the adventurous. Avalon is said to be a place of more freedom, but there's mention of a mysterious "them" who seem to be a controlling bunch, if only in the way of a posse of quasi-benevolent religious supervisors checking to see if you've passed some never-specified test. It's a land of peace, but then, so is the after-life. Is Avalon really the country from whose bourne no traveller returns, as we're led from time to time to believe? But then we're led to believe otherwise, because Alex has been there and has come back. No sooner does Bryher strew around a few literary and metaphysical allusions than she tramples them underfoot. This is maddening in the way of vampire novels that break the long-established rules ("What do you mean, you *like* garlic?"). Sometimes it's as if Bryher forgot exactly what it was she'd set out to do.

Visa for Avalon purports to take place in the future—television and computers are mentioned in passing—but the foreground is taken up by a great deal of linoleum: good, bright linoleum, and bad, squalid linoleum, and even a linoleum hat. Nothing dates an era more tellingly than floor coverings, and 1965 was the Age of Shag Carpeting, not the Age of Linoleum. In the physical details of its setting, *Visa for Avalon* suggests not a future but a past—two decades of it laminated together. There's the utopian brutalism of the fascist and Nazi 1930s ideologies, with their impulse to destroy the systems of the past and streamline the present, and there's the dinginess of the war years in England, with their crowded trains and depressing waiting rooms. The emotional climate, too, is that of wartime: the inability to get anywhere or obtain much-needed documents or do anything effective, combined with a grinding boredom, and, at the same time, the acute, stomach-churning anxiety of not knowing what is really going on.

In these areas of observation, *Visa for Avalon* has the texture of lived experience. We don't know what to think about the tyrannical government or Movement that's taking over—are they left or right, or does it matter?—but we certainly learn to the last traffic jam and hastily packed suitcase and nasty armband-wearing guard and eerily deserted street how such a takeover would feel to ordinary folk trapped inside it. As people do when their adrenaline levels are high and there's no overt means of expression, the characters focus on single details perceived with hallucinatory clearness: the rusty oil drum, the splintered piece of timber, the cable wheel. The handling of these sections is realistic in the extreme.

Once on the plane with the small saving remnant, we find ourselves back in a quasi-symbolic universe. Robinson wonders whether the whole experience he's just lived through has been an illusion; he decides it's real, but we're wondering. Soon he's speaking the language of salvation by Grace: "What had he ever done that had made him worthy of rescue?"

"Things do come to an end," says Lilian. As she remembers with nostalgia her life in Trelawney, she has an amazing insight:

> By a terrific effort of will and with the physical force that she had needed to stand against the autumn gales, she stammered, but who heard her against the roar of the engines, "I wanted to be out on the Seven Seas, I never wanted to be in Rose Cottage at all!"

It now seems that Lilian—unknown to herself—has wanted to run away to sea all along, like the young Bryher. Is a visa for Avalon a kind of litmus paper that shows us the truth about ourselves?

Right after this surprising *cri du coeur* the plane plunges into fog, the radio fails, the novel takes us through a near-death experience, and Robinson is back in Tennysonian mode: "All of us have our fate . . . none of us can escape it." But whatever his fate may be, it doesn't include a crash-landing in the sea, since Avalon itself is glimpsed briefly at the book's end: ". . . the clouds parted for an instant and Robinson saw far below them, as they came in for a perfect landing, gorse bushes, the valley full of apple trees and a stretch of white sand." Or, as Tennyson had it,

> I am going a long way
> With these thou seest—if indeed I go
> For all my mind is clouded with a doubt
> To the island-valley of Avilion;
> . . . it lies
> Deep-meadow'd, happy, fair, with orchard lawns . . .

Is life a waiting room or a journey? In *Visa for Avalon*, both options are proposed. If a waiting room, what comes after you've done the waiting? If a journey, what is its end? Bryher doesn't tell us, partly—one feels—because Tennyson doesn't tell us whether Arthur lived or died, and partly because she hadn't made up her own mind about life after death; but partly, also, because she wisely judged that in a narrative such as this, to travel is much better than to arrive.

There's some suggestion that Avalon is whatever you think it is, and the same can be said of *Visa for Avalon*. In part, it's a trip through the nightmare of political repression and mob takeover, in part a veiled encounter with approaching death: *Everyman* meets *The Pilgrim's Progress* crossed with "The Passing of Arthur" with undertones of *The Seventh Seal*, as domesticated in Trelawney-by-the-Sea. It would be stretching matters to call it an entirely successful work of art—its threads are too loose—but, despite this, it remains a suggestive and beguiling fiction by one of the twentieth century's most interesting artistic figures. The Paris Press should be thanked for republishing it.

Brave New World
by Aldous Huxley

O brave new world, that has such people in't!
MIRANDA, in Shakespeare's play *The Tempest*,
on first sighting the shipwrecked courtiers

In the latter half of the twentieth century, two visionary books cast their shadows over our futures. One was George Orwell's 1949 novel, *Nineteen Eighty-Four*, with its horrific vision of a brutal, mind-controlling totalitarian state—a book that gave us Big Brother, and thoughtcrime and Newspeak and the memory hole and the torture palace called the Ministry of Love, and the discouraging spectacle of a boot grinding into the human face forever.

The other was Aldous Huxley's *Brave New World* (1932), which proposed a different and softer form of totalitarianism—one of conformity achieved through engineered, bottle-grown babies and hypnotic persuasion rather than through brutality, of boundless consumption that keeps the wheels of production turning and of officially enforced promiscuity that does away with sexual frustration, of a pre-ordained caste system ranging from a highly intelligent managerial class to a subgroup of dim-witted serfs programmed to love their menial work, and of soma, a drug that confers instant bliss with no side effects.

Which template would win? we wondered. During the Cold War, *Nineteen Eighty-Four* seemed to have the edge. But when the Berlin Wall fell in 1989, pundits proclaimed the end of history, shopping reigned triumphant, and there was already lots of quasi-soma percolating through society. True, promiscuity had taken a hit from AIDS, but on balance we seemed to be in for a trivial, giggly, drug-enhanced Spend-O-Rama: *Brave New World* was winning the race.

That picture changed too, with the attack on New York City's Twin Towers in 2001. Thoughtcrime and the boot grinding into the human face could not be got rid of so easily after all. The Ministry of Love is back with us, it appears, though it's no longer limited to the lands behind the former Iron Curtain: the West has its own versions now.

On the other hand, *Brave New World* hasn't gone away. Shopping malls stretch as far as the bulldozer can see. On the wilder fringes of the genetic engineering community, there are true believers prattling of the GenRich and the GenPoor—Huxley's Alphas and Epsilons—and busily engaging in schemes for genetic enhancement and—to go *Brave New World* one better—for immortality.

Would it be possible for both of these futures— the hard and the soft—to exist at the same time, in the same place? And what would that be like?

Surely it's time to look again at *Brave New World* and to examine its arguments for and against the totally planned society it describes, in which "everybody is happy now." What sort of happiness is on offer, and what is the price we might pay to achieve it?

. . . . .

I first read *Brave New World* in the early 1950s, when I was fourteen. It made a deep impression on me, though I didn't fully understand some of what I was reading. It's a tribute to Huxley's writing skills that although I didn't know what knickers were, or camisoles—nor did I know that zippers, when they first appeared, had been denounced from pulpits as lures of the Devil because they made clothes so easy to take off—I nonetheless had a vivid picture of "zippicamiknicks," that female undergarment with a single zipper down the front that could be shucked so very easily:

> Zip! The rounded pinkness fell apart like a neatly divided apple. A wriggle of the arms, a lifting first of the right foot, then the left: the zippicamiknicks were lying lifeless and as though deflated on the floor.

I myself was living in the era of "elasticized panty girdles" that could not be got out of or indeed into without an epic struggle, so this was heady stuff indeed.

The girl shedding the zippicamiknicks is Lenina Crowne, a blue-eyed beauty both strangely innocent and alluringly voluptuous—or "pneumatic," as her many male admirers call her. Lenina doesn't see why she shouldn't have sex with anyone she likes whenever the occasion offers, as to do so is merely polite behaviour and not to do so is selfish. The man she's trying to seduce by shedding her undergarment is John "the Savage," who's been raised far outside the "civilized" pale on a diet of Shakespeare's chastity/whore speeches, and Zuni cults, and self-flagellation, and who believes in religion and romance, and in suffering to be worthy of one's beloved, and who idolizes Lenina until she doffs her zippicamiknicks in such a casual and shameless fashion.

Never were two sets of desiring genitalia so thoroughly at odds. And thereby hangs Huxley's tale.

. . . . .

Brave New World is either a perfect-world utopia or its nasty opposite, a dystopia, depending on your point of view: its inhabitants are beautiful, secure, and free from diseases and worries, though in a way we like to think we would find unacceptable. "Utopia" is sometimes said to mean "no place," from the Greek "O Topia," but others derive it from "eu," as in "eugenics," in which case it would mean "healthy place" or "good place." Sir Thomas More, in his own sixteenth-century *Utopia*, may have been punning: utopia is the good place that doesn't exist.

As a literary construct, *Brave New World* thus has a long list of literary ancestors. Plato's *Republic* and the Bible's Book of Revelation and the myth of Atlantis are the great-great-grandparents of the form; nearer in time are Sir Thomas More's *Utopia*, and the land of the talking-horse, totally rational Houyhnhnms in Jonathan Swift's *Gulliver's Travels*, and H. G. Wells's *The Time Machine*, in which the brainless, pretty "upper classes" play in the sunshine during the day and the ugly "lower classes" run the underground machinery and emerge at night to eat the social butterflies. . . .

Insofar as they are critical of society as it presently exists but nevertheless take a dim view of the prospects of the human race, utopias may verge on satire; but insofar as they endorse the view that humanity is perfectible, or can at least be vastly improved, they will resemble idealizing romances. The First World War marked the end of the romantic-idealistic utopian dream in literature, just as several real-life utopian plans were about to be launched with disastrous effects. The Communist regime in Russia and the Nazi takeover of Germany both began as utopian visions.

But as most literary utopias had already discovered, perfectibility breaks on the rock of dissent. What do you do with people who don't endorse your views or fit in with your plans? . . . It's rats in the eyes for you—as in *Nineteen Eighty-Four*—if you won't love Big Brother. (*Brave New World* has its own gentler punishments: for non-conformists, it's exile to Iceland, where Man's Final End can be discussed among like-minded intellects, without pestering "normal" people—in a sort of university, as it were.)

Utopias and dystopias from Plato's *Republic* on have had to cover the same basic ground that real societies do. All must answer the same questions: Where do people live, what do they eat, what do they wear, what do they do about sex and child-rearing? Who has the power, who does the work, how do citizens relate to nature, and how does the economy function? Romantic utopias such as Morris's *News from Nowhere* and W. H. Hudson's *A Crystal Age* present a Pre-Raphaelite picture, with the inhabitants going in for flowing robes, natural settings in abodes that sound like English country houses with extra stained glass, and lots of arts and crafts. Everything would be fine, we're told, if we could only do away with industrialism and get back in tune with Nature, and deal with overpopulation. . . .

But when Huxley was writing *Brave New World* at the beginning of the 1930s, he was, in his own words, an "amused, Pyrrhonic aesthete," a member of that group of bright young upstarts that swirled around the Bloomsbury Group and delighted in attacking anything Victorian or Edwardian. So *Brave New World* tosses out the flowing robes, the crafts, and the tree-hugging. Its architecture is futuristic—electrically lighted towers and softly glowing pink glass—and everything in its cityscape is relentlessly unnatural and just as relentlessly industrialized. Viscose and acetate and imitation leather are its fabrics of choice; apartment buildings, complete with artificial music and taps that flow with perfume, are its dwellings; transportation is by private helicopter. Babies are no longer born, they're grown in hatcheries, their bottles moving along assembly lines, in various types and batches according to the needs of "the hive," and fed on "external secretion" rather than "milk." The word *mother*—so thoroughly worshipped by the Victorians—has become a shocking obscenity; and indiscriminate sex, which was a shocking obscenity for the Victorians, is now *de rigueur.*

> "He patted me on the behind this afternoon," said Lenina.
>
> "There, you see!" Fanny was triumphant. "That shows what *he* stands for. The strictest conventionality."

Many of *Brave New World*'s nervous jokes turn on these kinds of inversions—more startling to its first audience, perhaps, than to us, but still wry enough. Victorian thrift turns into the obligation to spend, Victorian till-death-do-us-part monogamy has been replaced with "everyone belongs to everyone else," Victorian religiosity has been channelled into the worship of an invented deity—"Our Ford," named after American car-czar Henry Ford, god of the assembly line—via communal orgies. Even the "Our Ford" chant—"orgy-porgy"—is an inversion of the familiar nursery rhyme in which kissing the girls makes them cry. Now, it's if you refuse to kiss them—as "the Savage" does—that the tears will flow.

Sex is often centre stage in utopias and dystopias—who can do what, with which set of genital organs, and with whom being one of humanity's main preoccupations. Because sex and procreation have been separated and women no longer give birth—the very idea is yuck-making to them—sex has become a recreation. Little naked children carry on "erotic play" in the shrubberies, so as to get a hand in early. Some women are sterile—"freemartins"—and perfectly nice girls, though a little whiskery. The others practise "Malthusian drill"—a form of birth control—and take "pregnancy surrogate" hormone treatments if they feel broody, and sport sweet little faux-leather fashionista cartridge belts crammed with contraceptives. If they slip up on their Malthusian drill, there's always the lovely pink-glass Abortion Centre. Huxley wrote before the pill, but its advent brought his imagined sexual free-for-all a few steps closer. (What about gays? Does "everyone belongs to everyone else" really mean *everyone*? We aren't told.)

Of course, Huxley himself still had one foot in the nineteenth century: he could not have dreamed his upside-down morality unless he himself also found it threatening. At the time he was writing *Brave New World*, he was still in shock from a visit to the United States, where he was particularly frightened by mass consumerism and its group mentality and its vulgarities.

I use the word *dreamed* advisedly because *Brave New World*—gulped down whole—achieves an effect not unlike a controlled hallucination. All is surface; there is no depth. As you might expect from an author with impaired eyesight, the visual sense predominates: colours are intense, light and darkness vividly described. Sound is next in importance, especially during group ceremonies and orgies, and the viewing of "Feelies"—movies in which you feel the sensations of those onscreen, "The Gorillas' Wedding" and "Sperm Whale's Love-Life" being sample titles. Scents are third—perfume wafts everywhere, and is dabbed here and there; one of the most poignant encounters between John the Savage and the lovely Lenina is the one in which he buries his worshipping face in her divinely scented undergarments while she herself is innocently sleeping, zonked out on a strong dose of soma, partly because she can't stand the awful real-life smells of the "reservation" where the new world has not been implemented.

Many utopias and dystopias emphasize food (delicious or awful; or, in the case of Swift's Houyhnhnms, oats), but in *Brave New World* the menus are not presented. Lenina and her lay-of-the-month, Henry, eat "an excellent meal," but we aren't told what it is. (Beef would be my guess, in view of the huge barns full of cows that provide the external secretions.) Despite the dollops of sex-on-demand, the bodies in *Brave New World* are oddly disembodied, which serves to underscore one of Huxley's points: in a world in which everything is available, nothing has any meaning.

Meaning has in fact been eliminated, as far as possible. All books except works of technology have been banned, *pace* Ray Bradbury's 1953 novel *Fahrenheit 451*; museum-goers have been slaughtered, *pace* Henry Ford's "History is bunk." As for God, he is present "as an absence; as though he weren't there at all"—except, of course, for the deeply religious John the Savage, who has been raised on a Zuni "reservation" off-limits to normal Brave New Worlders. There, archaic life carries on, replete with "meaning" of the most intense kinds. John is the only character in the book who has a real body, but he knows it through pain, not through pleasure. "Nothing costs enough here," he says of the perfumed new world where he's been brought as an "experiment."

The "comfort" offered by Mustapha Mond—one of the ten

"Controllers" of this world and a direct descendant of Plato's Guardians—is not enough for John. He wants the old world back—dirt, diseases, free will, fear, anguish, blood, sweat, tears, and all. He believes he has a soul, and like many an early twentieth-century literary possessor of such a thing—such as the missionary in Somerset Maugham's 1921 story "Miss Thompson," who hangs himself after sinning with a prostitute—John is made to pay the price for this belief.

In the Foreword to *Brave New World* written in 1946, after the horrors of the Second World War and Hitler's Final Solution, Huxley criticizes himself for having provided only two choices in his 1932 utopia/dystopia—an "insane life in Utopia" or "the life of a primitive in an Indian village, more human in some respects, but in others hardly less queer and abnormal." (He does, in fact, provide a third sort of life—that of the intellectual community of misfits on Iceland—but poor John the Savage isn't allowed to go there, and he wouldn't have liked it anyway, as there are no public flagellations available.) The Huxley of 1946 comes up with another sort of utopia, one in which "sanity" is possible. By this, Huxley means a kind of "High Utilitarianism" dedicated to a "conscious and rational" pursuit of man's "Final End," which is a kind of union with the immanent "Tao or Logos, the transcendent Godhead or Brahmin." No wonder Huxley subsequently got heavily into the mescaline and wrote *The Doors of Perception*, thus inspiring a generation of 1960s hopheads and musicians to seek God in altered brain chemistry. His interest in soma, it appears, didn't spring out of nowhere.

Meanwhile, those of us still tottering along on the earthly plane—and thus still able to read books—are left with *Brave New World*. How does it stand up, seventy-five years later? And how close have we come, in real life, to the society of vapid consumers, idle pleasure-seekers, inner-space trippers, and programmed conformists that it presents?

The answer to the first question, for me, is that it stands up very well. It's still as vibrant, fresh, and somehow shocking as it was when I, for one, first read it.

The answer to the second question, Dear Reader, rests with you. Look in the mirror: do you see Lenina Crowne looking back at you or do you see John the Savage? If you're a human being, you'll be seeing something of both, because we've always wanted things both ways. We wish to be as the careless gods, lying around on Olympus, eternally beautiful, having sex and being entertained by the anguish of others. And at the same time we want to be those anguished others, because we believe, with John, that life has meaning beyond the play of the senses and that immediate gratification will never be enough.

It was Huxley's genius to present us to ourselves in all our ambiguity. Alone among the animals, we suffer from the future perfect tense. Rover the Dog cannot imagine a future world of dogs in which all fleas will have been eliminated and doghood will finally have achieved its full glorious potential. But thanks to their uniquely structured languages, human beings can imagine such enhanced states for themselves, though they can also question their own grandiose constructions. It's these double-sided abilities that produce masterpieces of speculation such as *Brave New World*.

To quote *The Tempest*, source of Huxley's title: "We are such stuff/As dreams are made on." He might well have added: and nightmares.

Of the Madness of Mad Scientists: Jonathan Swift's Grand Academy

In the late 1950s, when I was a university student, there were still B movies. They were inexpensively made and lurid in nature, and you could see them at cheap matinee double bills as a means of escaping from your studies. Alien invasions, mind-altering potions, and scientific experiments gone awry featured largely.

Mad scientists were a staple of the B-film double bill. Presented with a clutch of white-coated men wielding test tubes, we viewers knew at once—being children of our times—that at least one of them would prove to be a cunning megalomaniac bent on taking over the world, all the while subjecting blondes to horrific experiments from which only the male lead could rescue them, though not before the mad scientist had revealed his true nature by gibbering and raving. Occasionally the scientists were lone heroes, fighting epidemics and defying superstitious mobs bent on opposing the truth by pulverizing the scientist, but the more usual model was the lunatic. When the scientists weren't crazy, they were deluded: their well-meaning inventions were doomed to run out of control, creating havoc, tumult, and piles of messy guck, until gunned down or exploded just before the end of the film.

. . . . .

Where did the mad scientist stock figure come from? How did the scientist—the imagined kind—become so very deluded and/or demented?

It wasn't always like that. Once upon a time there weren't any scientists as such, in plays or fictions, because there wasn't any science as such, or not science as we know it today. There were alchemists and dabblers in black magic—sometimes one and the same—and they were depicted not as lunatics but as charlatans bent on fleecing the unwary by promising to turn lead into gold, or else as wicked pact-makers with the Devil, hoping—like Dr. Faustus—to gain worldly wealth, knowledge, and power in exchange for their souls. The too-clever-by-half part of their characters may have descended from Plato's Atlanteans or the builders of Babel—ambitious exceeders of the boundaries set for human beings, usually by some god, and destroyed for their presumption. These alchemists and Faustian magicians certainly form part of the mad scientist's ancestral lineage, but they aren't crazy or deluded, just daring and immoral.

It's a considerable leap from them to the excesses of the wild-eyed B-movie scientists. There must be a missing link somewhere, like the walking seal discovered just recently—though postulated by Charles Darwin as a link between a walking canid and a swimming seal. For the mad scientist missing link, I propose Jonathan Swift, acting in synergy with the Royal Society. Without the Royal Society, no *Gulliver's Travels*, or not one with scientists in it; without *Gulliver's Travels*, no mad scientists in books and films. So goes my theory.

. . . . .

I read Jonathan Swift's *Gulliver's Travels* as a child, before I knew anything about the B-movie scientists. Nobody told me to read it; on the other hand, nobody told me not to. The edition I had was not a child's version, of the kind that dwells on the cute little people and the funny giant people and the talking horses, but dodges any mention of nipples and urination, and downplays the excrement. These truncated versions also leave out most of Part Three—the floating island of Laputa, the Grand Academy of Lagado with its five hundred scientific experiments, and the immortal Struldbrugs of Luggnagg—as being incomprehensible to young minds. My edition was unabridged, and I didn't skip any of it, Part Three included. I read the whole thing.

I thought it was pretty good. I didn't yet know that *Gulliver's Travels* was satirical, that Mr. Swift's tongue had been rammed very firmly into his cheek while writing it, and that even the name "Gulliver," so close to "gullible," was a tip-off. I believed the letters printed at the beginning—the one from Mr. Gulliver himself, complaining about the shoddy way in which his book had been published, and the one from his cousin Mr. Sympson—so close to "simpleton," I later realized—testifying to the truthfulness of Mr. Gulliver. I did understand that someone called Mr. Swift had had something to do with this book, but I didn't think he'd just made all of it up. In early eighteenth-century terms, the book was a "bite"—a tall tale presented as the straight-faced truth in order to sucker the listener into believing it—and I got bitten.

Thus I first read this book in a practical and straightforward way, much in the way it is written. For instance, when Mr. Gulliver pissed on the fire in the royal Lilliputian palace in order to put it out, I didn't find this either a potentially seditious poke at the pretensions of royalty and the unfairness of courts or a hilarious vulgarism. Rather, having been trained myself in the time-honoured woodsman's ways of putting out campfires, I thought Mr. Gulliver had displayed an admirable presence of mind.

The miniature people and the giants did hint to me of fairy tales, but Part Three—the floating island and the scientific establishment—didn't seem to me all that far-fetched. I was then living in what was still the golden or bug-eyed monster age of science fiction—the late 1940s—so I took spaceships for granted. This was before the disappointing news had come in—no intelligent life on Mars—and also before I'd read H. G. Wells's *The War of the Worlds*, in the light of which any life intelligent enough to build spaceships and come to Earth would be so much smarter than us that we'd be viewed by them as ambulatory kebabs. So I considered it entirely possible that, once I'd grown up, I might fly through space and meet some extraterrestrials, who then as now were considered to be bald, with very large eyes and heads.

Why then couldn't there be a flying island such as Laputa? I thought the method of keeping the thing afloat with magnets was a little cumbersome—hadn't Mr. Swift heard of jet propulsion?—but the idea of hovering over a country that was annoying you so they'd be in full shadow and their crops wouldn't grow seemed quite smart. As for dropping stones onto them, it made perfect sense: kids of the immediately postwar generation were well versed in the advisability of air superiority, and knew a lot about bombers.

I didn't understand why these floating-island people had to eat food cut into the shapes of musical instruments, but the flappers who hit them with inflated bladders to snap them out of their thought trances didn't seem out of the question. My father was by that time teaching in the Department of Zoology at the

University of Toronto, and growing up among the scientists, and thus being able to observe them at work, I knew they could be like that: the head of the Zoology Department at the University of Toronto was notorious for setting himself on fire by putting his still-smouldering pipe into his pocket and could have made excellent use of a flapper.

When I got as far as the Grand Academy of Lagado I felt right at home. In addition to being the golden age of bug-eyed monsters, the late 1940s was also the golden age of dangerous chemistry sets for children—now prohibited, no doubt wisely—and my brother had one. "Turn water to blood and astonish your friends!" proclaimed the advertisements, and this was no sooner said than done, with the aid of a desirable crystal named—as I recall—potassium permanganate. There were many other ways in which we could astonish our friends, and short of poisoning them, we did all of them. I doubt that we were the only children to produce hydrogen sulphide ("Make the smell of rotten eggs and astonish your friends!") on the day when our mother's bridge club was scheduled to meet. Through these experiments, we learned the rudiments of the scientific method: any procedure done in the same way with the same materials ought to produce the same results. And ours did, until the potassium permanganate ran out.

These were not the only experiments we performed. I will not catalogue our other adventures in science, which had their casualties—the jars of tadpoles dead from being left by mistake in the sun, the caterpillars that came to sticky ends—but will pause briefly to note the mould experiment, consisting of various foodstuffs placed in jars—our home-preserving household had a useful supply of jars—to see what might grow on them in the way of mould. Many-coloured and whiskery were the results, which

I mention now only to explain why the Grand Academy "projector" who thought it might be a brilliant idea to inflate a dog through its nether orifice in order to cure it of colic raised neither of my eyebrows. It was a shame that the dog exploded, but this was surely a mistake in the method rather than a flaw in the concept; or that was my opinion.

Indeed, this scene stayed with me as a memory trace that was reactivated the first time I had a colonoscopy and was myself inflated in this way. You had the right idea, Mr. Swift, I mused, but the wrong application. Also, you thought you were being ridiculous. Had you known that the dog-enlarging anal bellows you must have found so amusing would actually appear on Earth two hundred and fifty years later in order to help doctors run a tiny camera through your intestines so they could see what was going on in there, what would you have said?

And so it is with the majority of the experiments described in the Grand Academy chapters of *Gulliver's Travels*. Swift thought them up as jokes, but many of them have since been done in earnest, though with a twist. For instance, the first "projector" Gulliver meets is a man who has run himself into poverty through the pursuit of what Swift devised as a nutty-professor chase-a-moonbeam concept: this man wants to extract sunbeams out of cucumbers so he can bottle them for use in the winter, when the supply of sunbeams is limited. Swift must have laughed into his sleeve, but I, the child reader, found nothing extraordinary in this idea because every morning I was given a spoonful of cod liver oil, bursting with vitamin D, the "Sunshine Vitamin." The projector had simply used the wrong object—cucumbers instead of cod.

Some of the experiments being done by the projectors interested me less, though they have since contributed to Swift's reputation for prescience. The blind man at the Academy who's teaching other blind people to distinguish colours by touch was doubtless intended by Swift to represent yet more foolishness on the part of would-be geniuses, but now there are ongoing experiments involving something called the BrainPort—a device designed to allow blind people to "see" with their tongues. The machine with many handles that, when turned, cause an array of oddly Chinese-looking words to arrange themselves into an endless number of sequences—thus writing masterpieces eventually, like the well-known infinitely large mob of monkeys with typewriters—is now thought by some to be a forerunner of the computer.

Predicting the future and suggesting the invention of handy new devices was, however, very far from Swift's intention. His "projectors"—so called because they are absorbed in their projects—are a combination of experimental scientist and entrepreneur; they exist within *Gulliver's Travels* as pearls on his long string of human folly and depravity, midway between the Lilliputians and their tiny fracas and petty intrigues and the brutal, nasty, smelly, ugly, and vicious Yahoos of the fourth book, who represent humanity in its bared-to-the-elements Hobbesian basic state.

But Swift's projectors aren't wicked, and they aren't really demented. They're even well meaning: their inventions are intended for the improvement of humankind. All we have to do is give them more money and more time and let them have their way, and everything will get a lot better very soon. It's a likely story, and one we've heard many times since the advent of applied science. Sometimes this story ends well, at least for a while—science did lower the human mortality rate, the automobile did speed up travel, air conditioning did make us cooler in summer, the "green revolution" did increase the supply of food. But the doctrine of unintended consequences applies quite regularly to the results of scientific "improvements": agriculture can't keep up with the population explosion with the result that millions are leading lives of poverty and misery, air conditioning contributes to global warming, the automobile promised freedom until—via long commute distances, clogged roads, and increased pollution—it delivered servitude. Swift anticipated us: the projectors promise an idyllic future in which one man shall do the work of ten and all fruits shall be available at all times—*pace* automation and the supermarket—but "The only inconvenience is, that none of these projects are yet brought to perfection, and in the meantime, the whole country lies miserably waste, the houses in ruins, and the people without food or clothes." Under the influence of the projectors the utopian pie is visible in the sky, but it remains there.

As I've said, the projectors are not intentionally wicked. But they have tunnel vision—much like a present-day scientist quoted recently, who, when asked why he'd created a polio virus from scratch, answered that he'd done it because the polio virus was a simple one, and that next time he'd create a more complex virus. A question most of us would have understood to have meant, "Why did you do such a potentially dangerous thing?"—a question about ends—was taken by him to be a question about means. Swift's projectors show the same confusion in their understanding of ordinary human desires and fears. Their greatest offence is not against morals: instead they are offenders against common sense—what Swift might have called merely "sense." They don't intend to cause harm, but by refusing to admit the adverse consequences of their actions, they cause it anyway.

The Grand Academy of Lagado was recognized by Swift's readers as a satire upon the Royal Society, which even by Swift's time was an august and respected institution. Though English seekers after empirical facts had been meeting since 1640, the group became formalized as the Royal Society under Charles II, and as of 1663 was referred to as "The Royal Society of London for Improving Natural Knowledge." The word *natural* signifies the distinction between such knowledge—based on what you could see and measure, and on the "scientific method": some combination of observation, hypothesis, deduction, and experiment—from "divine" knowledge, which was thought to be invisible and immeasurable, and of a higher order.

Though these two orders of knowledge were not supposed to be in conflict, they often were, and both kinds might be brought to bear on the same problem, with opposite results. This was especially true during outbreaks of disease: victims and their families would resort both to prayer and to purging, and who could tell which might be the more efficacious? But in the first fifty years of the Royal Society's existence, "natural knowledge" gained much ground, and the Royal Society acted increasingly as a peer-review body for experiments, fact-gathering, and demonstrations of many kinds.

Swift is thought to have begun *Gulliver's Travels* in 1721, which was interestingly enough the year in which a deadly smallpox epidemic broke out, both in London and in Boston, Massachusetts. There had been many such epidemics, but this one saw the eruption of a heated controversy over the practise of inoculation. Divine knowledge had varying views: was inoculation a gift from God, or was smallpox itself a divine visitation and punishment for misbehaviour, with any attempt made to interfere with it being impiety? But practical results rather than theological arguments were being increasingly credited.

In London, inoculation was championed by Lady Mary Wortley Montagu, who had learned of the practise in Turkey when her husband had been ambassador there; in Boston, its great supporter was, oddly enough, Cotton Mather—he of the Salem witchcraft craze and *The Wonders of the Invisible World*—who had been told of it by an inoculated slave from Africa. Both, though initially vilified, were ultimately successful in their efforts to vindicate the practise. Both acted in concert with medical doctors—Mather with Dr. Zabdiel Boylston, who, in 1726, read a paper on the results of his practise-cum-research to the Royal Society, Lady Mary with Dr. John Arbuthnot.

You might think Swift would have been opposed to inoculation. After all, the actual practise of inoculation was repulsive and counterintuitive, involving as it did the introduction of pus from festering victims into the tissues of healthy people. This sounds quite a lot like the exploding dog from the Grand Academy of Lagado and such other Lagadan follies. In fact, Swift took the part of the inoculators. He was an old friend of Dr. Arbuthnot, a fellow member of the Martinus Scriblerus Club of 1714, a group that had busied itself with satires on the abuses of learning. And, unlike the ridiculous experiments of the "projectors"—experiments that may have been invented by Swift with the aid of some insider hints from Dr. Arbuthnot—inoculation seemed actually to work, most of the time.

It isn't experimentation as such that's the target of Book Three, but experiments that backfire. Moreover, it's the obsessive nature of the projectors: no matter how many dogs they explode, they keep at it, certain that the next time they inflate a dog they'll achieve the proposed result. Although they appear to be acting according to the scientific method, they've got it backward. They think that because their reasoning tells them the experiment ought to work, they're on the right path; thus they ignore observed experience. Although they don't display the full-blown madness of the truly mad fictional scientists of the mid-twentieth century, they're a definitive step along the way: the Lagadan Grand Academy was the literary mutation that led to the crazed white-coats of those B movies.

. . . . .

There were many intermediary forms. Foremost among them was, of course, Mary Shelley's Dr. Frankenstein, he of the man-made monster—a good example of an obsessive scientist blind to all else as he seeks to prove his theories by creating a perfect man out of dead bodies. The first to suffer from his blindness and single-mindedness is his fiancée, murdered by the creature on Dr. Frankenstein's wedding night in revenge for Frankenstein's refusal to love and acknowledge the living being he himself has created. Next came Hawthorne's various obsessed experimenters. There's Dr. Rappaccini, who feeds poison in small amounts to his daughter, thus making her immune to it though she is poisonous to others and is thus cut off from life and love. There's also the "man of science" in "The Birthmark," who becomes fixated on the blood-coloured, hand-shaped birthmark of his beautiful wife. In an attempt to remove it through his science—thus rendering her perfect—he takes her to his mysterious laboratory and administers a potion that undoes the bonds holding spirit and flesh together, which kills her.

Both of these men—like Dr. Frankenstein—prefer their own arcane knowledge and the demonstration of their power to the safety and happiness of those whom they ought to love and cherish. In this way they are selfish and cold, much like the Lagadan projectors who stick to their theories no matter how much destruction and misery they may cause. And both, like Dr. Frankenstein, cross the boundaries set for human beings and dabble in matters that are either (a) better left to God or (b) none of their business.

The Lagadan projectors were both ridiculous and destructive, but in the middle of the nineteenth century the mad scientist line splits in two, with the ridiculous branch culminating in the Jerry Lewis "nutty professor" comic version and the other leading in a more tragic direction. Even in "alchemist" tales like the Faustus story, the comic potential was there—Faustus on the stage was a great practical joker—but in darker sagas like *Frankenstein* this vein is not exploited.

In modern times the "nutty professor" trope can probably trace its origins to Thomas Hughes's extraordinarily popular 1857 novel, *Tom Brown's School Days*. There we meet a boy called Martin, whose nickname is "Madman." Madman would rather do chemical experiments and explore biology than parse Latin sentences—a bent the author rather approves than not, as he sees in Madman the coming age:

> If we knew how to use our boys, Martin would have been seized upon and educated as a natural philosopher. He had a passion for birds, beasts, and insects, and knew more of them and their habits than any one in Rugby. . . . He was also an experimental chemist on a small scale, and had made unto himself an electric machine, from which it was his greatest pleasure and glory to administer small shocks to any small boys who were rash enough

> to venture into his study. And this was by no means an adventure free from excitement; for besides the probability of a snake dropping on to your head or twining lovingly up your leg, or a rat getting into your breeches-pocket in search of food, there was the animal and chemical odour to be faced, which always hung about the den, and the chance of being blown up in some of the many experiments which Martin was always trying, with the most wondrous results in the shape of explosions and smells that mortal boy ever heard of.

Despite the indulgent tone, the Lagadan comic aspects are in evidence: the chemical experiments that blow up, the stinky substances, the mess, the animal excrement, the obsession.

The tragic or sinister mad scientist evolutionary line runs through R. L. Stevenson's 1886 novel, *Dr. Jekyll and Mr. Hyde*, in which Dr. Jekyll—another of those cross-the-forbidden-liners, with another of those mysterious laboratories—stumbles upon, or possibly inherits from Hawthorne, another of those potions that dissolve the bonds holding spirit and flesh together. But this time the potion doesn't kill the drinker, or not at first. It does dissolve his flesh, but then it alters and re-forms both body and soul. There are now two selves, which share memory, but nothing else except the house keys. Jekyll's potion-induced second self, Hyde, is morally worse but physically stronger, with more pronounced "instincts." As this is a post-Darwinian fable, he is also hairier.

Dr. Jekyll is then betrayed by the very scientific method he has relied upon. Time after time, the mixing up of the potion and the drinking of it produce the same results; so far, so good-and-bad. But then the original supply of chemicals runs out, and the new batch doesn't work. The boundary-dissolving element is missing, and Dr. Jekyll is fatally trapped inside his furry, low-browed, murderous double. There were earlier "sinister double" stories, but this one—to my knowledge—is the first in which the doubling is produced by a "scientific" chemical catalyst. As with much else, this kind of transmutation has become a much-used comic book and filmic device. (The Hulk, for instance—the raging, berserk alter ego of reserved physicist Bruce Banner—came by his greenness and bulkiness through exposure to the rays from a "gamma bomb" trial created by Dr. Banner himself.)

Next in the line comes H. G. Wells's 1896 Dr. Moreau—he of the Island, upon which he attempts, through cruel vivisection experiments, to sculpt animals into people, with appalling and eventually lethal results. Moreau has lost the well-meaning but misguided quality of the projectors: he's possessed by a "passion for research" that exists for its own sake, simply to satisfy Moreau's own desire to explore the secrets of physiology. Like Frankenstein, he plays God—creating new beings—and the results are monstrous. And like so many of the sinister scientists who come after him, he is "irresponsible, so utterly careless! His curiosity, his mad, aimless investigations, drove him on. . . ."

From Moreau, it's a short step to the golden age of mad scientists, who became so numerous in both fiction and film by the mid-twentieth century that everyone recognized the stereotype as soon as it made its appearance.

Its lowest point is reached, quite possibly, in the B movie called variously *The Head That Wouldn't Die* or *The Brain That Wouldn't Die.* The scientist in it is even more seriously depraved than usual. The head in question is that of his girlfriend; it comes off in a car accident, after which incident most men might have cried. But the mad scientist is building a Frankenstein monster out of body parts filched from a hospital, underestimating as usual the monster's clothing size—why do those monsters' sleeves always end halfway down their arms?—so he wraps the girl's head in his coat and scampers off with it across the fields. Once under a glass bell with wires attached to its neck and its hair in a Bride of Frankenstein frizzle, the head gives itself to thoughts of revenge while the scientist himself haunts strip clubs in search of the perfect body to attach to it.

.....

There's another element in Book Three of *Gulliver's Travels* that bears mention here because it so often gets mixed into the alchemist/mad scientist sorts of tales: the theme of immortality. On the island of Luggnagg, the third in Swift's trio of capital-L islands, Gulliver encounters the immortals—children born with a spot on their foreheads that means they will never die. At first, Gulliver longs to meet these "Struldbrugs," whom he pictures as blessed: surely they will be repositories of knowledge and wisdom. But he soon finds that they are on the contrary cursed because, like their mythological forebears Tithonus and the Sibyl of Cumae, they do not receive eternal youth along with their eternal life. They simply live on and on, becoming older and older, and also "opinionated, peevish, covetous, morose, vain . . . and dead to all natural affection." Far from being envied, they are despised and hated; they long for death but cannot achieve it.

Immortality has been one of the constant desires of humanity. The means to it differ—one may receive it through natural means, as in Luggnagg, or from a god, or by drinking an elixir of life, or by passing through a mysterious fire, as in H. Rider Haggard's novel *She*, or by drinking the blood of a vampire; but there's always a dark side to it.

Luggnagg is Gulliver's last noteworthy Book Three stop. Through his encounter with the Struldbrugs, he's drawing close to the heart of Swift's matter: what it is to be human. In Book Four he plunges all the way in: his final voyage takes him to the land of the rational and moral talking-horse Houyhnhnms and brings him face-to-face with an astonishingly Darwinian view of humanity's essence. The filthy apelike beasts called Yahoos he encounters there are viewed by the Houyhnhnms as beasts, and treated as such; and, much to Gulliver's dismay, he is at last forced to recognize that, apart from a few superficial differences such as clothing and language, he, too, is a Yahoo.

As Swift's friend Alexander Pope wrote shortly after the publication of *Gulliver's Travels*: "The proper study of Mankind is Man." In our own age, that study is not only proper, it's more necessary than ever. The botched experiments of Swift's projectors and our own exponentially successful scientific discoveries and inventions are both driven by the same forces: human curiosity and human fears and desires. Since, increasingly, whatever we can imagine we can also enact, it's crucial that we understand what impels us. The mad scientist figure is—to paraphrase Oscar Wilde—our own Caliban's face in the mirror. Are we merely very smart Yahoos, and, if so, will we ultimately destroy ourselves and much else through our own inventions?

Science was just coming into being in the age of Swift. Now it's fully formed, but we're still afraid of it. Partly we fear its Moreau-like coldness, a coldness that is in fact real, for science as such does not have emotions or a system of morality built into it, any more than a toaster does. It's a tool—a tool for actualizing what we desire and defending against what we fear—and like any other tool, it can be used for good or ill. You can build a house with a hammer, and you can use the same hammer to murder your neighbour.

Human tool-makers always make tools that will help us get what we want, and what we want hasn't changed for thousands of years because as far as we can tell the human template hasn't changed either. We still want the purse that will always be filled with gold, and the Fountain of Youth. We want the table that will cover itself with delicious food whenever we say the word, and that will be cleaned up afterwards by invisible servants. We want the Seven-League Boots so we can travel very quickly, and the Hat of Darkness so we can snoop on other people without being seen. We want the weapon that will never miss, and the castle that will keep us safe. We want excitement and adventure; we want routine and security. We want to have a large number of sexually attractive partners, and we also want those we love to love us in return, and to be utterly faithful to us. We want cute, smart children who will treat us with the respect we deserve. We want to be surrounded by music, and by ravishing scents and attractive visual objects. We don't want to be too hot or too cold. We want to dance. We want to speak with the animals. We want to be envied. We want to be immortal. We want to be as gods.

But in addition, we want wisdom and justice. We want hope. We want to be good. Therefore we tell ourselves warning stories that deal with the shadow side of our other wants. Swift's Grand Academy and its projectors, and their descendants the mad scientists, are among those shadows.

. . . . .

Last week I came across a "project" that's a blend of art object and scientific experiment. Suspended in a glass bubble with wires attached to it—something straight out of a 1950s B movie, you'd think—is a strangely eighteenth-century Lilliputian coat. It's made of "Victimless Leather"—leather made of animal cells growing on a matrix. This leather is "victimless" because it has never been part of a living animal's skin. Yet the tiny coat is alive—or is it? What do we mean by "alive"? Can the experiment be terminated without causing "death"? Heated debates on this subject proliferate on the Internet.

The debate would have been right at home in Swift's Grand Academy: a clever but absurd object that's presented straight but is also a joke; yet not quite a joke, for it forces us to examine our preconceptions about the nature of biological life. Above all, like Swift's exploding dog and the proposal to extract sunshine out of cucumbers, the Victimless Leather garment is a complex creative exercise. If "What is it to be human?" is the central question of *Gulliver's Travels*, the ability to write such a book is itself part of the answer. We are not only what we do, we are also what we imagine. Perhaps, by imagining mad scientists and then letting them do their worst within the boundaries of our fictions, we hope to keep the real ones sane.

Five Tributes

An Introductory Note

SF is not only something I've written about, it's something I've written. I've discussed my three full-length forays into ustopia-writing in the chapter called "Dire Cartographies," but sprinkled here and there throughout my work, like breadcrumbs in the tangled wood, are a number of smaller homages to the various SF forms.

I've selected five of these: "Cryogenics" is a dinner-table conversation about getting your head frozen—a motif that appears not only in the review of Bill McKibben's *Enough* but also in *Oryx and Crake*, published—weirdly enough—at the same time. In "Cold-Blooded," the Earth is visited by aliens that happen to be giant insects; in "Homelanding," the aliens are being shown around by other aliens, who turn out to be us. "Time Capsule Found on the Dead Planet" is a riff on the time capsule theme so familiar to aficionados of classic SF. Finally, "The Peach Women of Aa'A" is an excerpt from *The Blind Assassin*—a spontaneous oral concoction presented by its male romantic lead to his lover in response to her demand for happy endings.

There were many other possible choices—the Lizard Men of Xenor, the attack of the Giant Sponge, the singing contest for world leaders, the Sasquatch, Man, and Two Androids . . . but some of the breadcrumbs should always remain in the forest.

Cryogenics: A Symposium

A. When I'm sixty-five I'm going to get my head cut off and flash-frozen. They've already got the technology, they've set up the corporations . . . Then it'll stay frozen until they've learned how to clone the rest of my body from a single cell, and they'll thaw out my head and reattach it. By that time, I figure the environment and all that stuff will be through the downturn and things will be more straightened out.

D. More Pinot Grigio? An olive?

A. Thanks. Some people are doing the whole body, but right now all I can afford is the head.

C. Ah. Market forces at work.

B. I take it you think your mind will survive this process, memories intact?

A. That would be the idea. Information storage, then retrieval at a later date . . .

B. Mind, or brain? Some people think the two are not coextensive. For instance, your brain might be a sort of grey Tastee-Freez, while your mind . . .

C. How about freezer burn? Ever seen frozen eyes? They go the colour of . . .

D. Would your new body be sixty-five too?

E. This Chilean sea bass is yummy!

B. We shouldn't be eating it. They're wiping it out. They are actually strip-mining the entire ocean. They're aiming for a huge underwater golf course.

D. I know, I know, but I forgot, and anyway it's already cooked so we might as well.

B. I was thinking more like twenty-three, for the body.

C. So you're going to have this wrinkly old head on top of a beef-cake? Not very delectable.

D. I wouldn't want to climb in the sack with something like that!

A. You won't be around, honey-bunny. Anyway they'll do plastic surgery. I'll look great. But I'll get to keep the wisdom I'll have accumulated by then.

E. You are *such* a dreamer! This whole thing is sooo grotesque!

A. New scientific ideas always seem grotesque to the masses.

E. I am *not* the *masses*! Anyway, what would stop them from taking your money, then after a few years with your head in the freezer they'd declare bankruptcy, pull out the plug, and toss your head in the garbage? That's what they'll do!

A. No need to be rude. I have faith in the process.

C. I've got a worse idea! They unfreeze your head and hook it up to a monitor, and run your most painful memories on it as cheap entertainment. Your whole life would be as a sideshow freak!

E. Or there would be a natural catastrophe—an earthquake, a tornado—the grid goes down—your head rots . . . Could you pass the slave-worker poison-sprayed artificially ripened grapes, please, and yes, I know I shouldn't have bought them. I did wash them, though. So don't worry.

A. I've thought of that. They'll have solar panels, with the lines running down into a shockproof underground cavern . . .

B. Look, let's face it. Pollution, vanishing ozone layer, genetically engineered organisms go on the rampage, the icebergs melt, the sea floods all coastal plains, plagues wipe out civilization . . . Only a few survive, reduced to roaming bands of brutal scavengers. They travel at night to avoid the deadly rays of the sun, and, all large land mammals having gone extinct, they eat rats, cockroaches, roots, and one another.

A. I'll be sleeping out that part, remember?

B. Wait . . . they come upon the underground cavern. There's no guards anymore and the hinges have rusted off the door. The nomads break in, they pry open the fridge, and what do they see?

D. A wedge of leftover Brie, half a head of celery, a thing of yogurt *way* past the sell-by date . . . Let's have coffee. This is shade-grown coffee, so don't look at me like that. Oh yeah, they also find that damn pike you caught last summer, sweetie, it's stinking up the entire freezer, what exactly are your plans for it?

B. Don't be frivolous. This is about his head. They open the freezer, and they see . . .

C. I think I know where this is going.

B. They see protein! They say, Get the cooking pot. They say, Feast time!

A. You are a pathetic, sick, psychically damaged individual.

B. I'm just a realist.

C. Same thing.

Cold-Blooded

To my sisters, the Iridescent Ones, the Egg-Bearers, the Many-Faceted, greetings from the Planet of Moths.

At last we have succeeded in establishing contact with the creatures here who, in their ability to communicate, to live in colonies, and to construct technologies, most resemble us, although in these particulars they have not advanced above a rudimentary level.

During our first observation of these "blood creatures," as we have termed them—after the colourful red liquid that is to be found inside their bodies, and that appears to be of great significance to them in their poems, wars, and religious rituals—we supposed them incapable of speech, as those specimens we were able to examine entirely lacked the organs for it. They had no wing-casings with which to stridulate—indeed they had no wings; they had no mandibles to click; and the chemical method was unknown to them, since they were devoid of antennae. "Smell," for them, is a perfunctory affair, confined to a flattened and numbed appendage on the front of the head. But after a time, we discovered that the incoherent squeakings and gruntings that emerged from them, especially when pinched, were in fact a form of language, and after that we made rapid progress.

We soon ascertained that their planet, named by us the Planet of Moths after its most prolific and noteworthy genus, is called by these creatures *Earth*. They have some notion that their ancestors were created from this substance; or so it is claimed in many of their charming but irrational folk tales.

In an attempt to establish common ground, we asked them at what season they mated with and then devoured their males. Imagine our embarrassment when we discovered that those individuals with whom we were conversing *were* males! (It is very hard to tell the difference, as their males are not diminutive, as ours are, but, if anything, bigger. Also, lacking natural beauty—brilliantly patterned carapaces, diaphanous wings, luminescent eyes, and the like—they attempt to imitate our kind by placing upon their bodies various multicoloured draperies, which conceal their generative parts.)

We apologized for our faux pas and inquired as to their own sexual practises. Picture our nausea and disgust when we discovered that it is the male, not the egg-bearer, which is the most prized among them! Abnormal as this will seem to you, my sisters, their leaders are for the most part male, which may account for their state of relative barbarism. Another peculiarity that must be noted is that, although they frequently kill them in many other ways, they rarely devour their females after procreation. This is a waste of protein; but then, they are a wasteful people.

We hastily abandoned this painful subject.

Next we asked them when they pupated. Here again, as in the case of "clothing"—the draperies we have mentioned—we uncovered a fumbling attempt at imitation of our kind. At some indeterminate point in their life cycles, they cause themselves to be placed in artificial stone or wooden cocoons, or chrysalises. They have an idea that they will someday emerge from these in an altered state, which they symbolize with carvings of themselves with wings. However, we did not observe that any had actually done so.

It is well to mention at this juncture that, in addition to the many species of moths for which it is justly famous among us, the Planet of Moths abounds in thousands of varieties of creatures that resemble our own distant ancestors. It seems that one of our previous attempts at colonization—an attempt so distant that our record of it is lost—must have borne fruit. However, these beings, although numerous and ingenious, are small in size and primitive in their social organization, and attempts to communicate with them were not—or have not been, so far—very successful. The blood-creatures are hostile toward them, and employ against them many poisonous sprays, traps, and so forth, in addition to a sinister manual device termed a "fly swatter." It is agonizing indeed to watch one of these instruments of torture and death being wielded by the large and frenzied against the small and helpless; but the rules of diplomacy forbid our intervention. (Luckily the blood-creatures cannot understand what we say to one another about them in our own language.)

But despite all the machinery of destruction that is aimed at them, our distant relatives are more than holding their own. They feed on the crops and herd-animals and even on the flesh of the blood-creatures; they live in their homes, devour their clothes, hide and flourish in the very cracks of their floors. When the blood-creatures have succeeded at last in overbreeding themselves, as it seems their intention to do, or in exterminating one another, rest assured that our kind, already superior in both numbers and adaptability, will be poised to achieve the ascendancy that is ours by natural right.

This will not happen tomorrow, but it will happen. As you know, my sisters, we have long been a patient race.

Homelanding

1.

Where should I begin? After all, you have never been there; or if you have, you may not have understood the significance of what you saw, or thought you saw. A window is a window, but there is looking out and looking in. The native you glimpsed, disappearing behind the curtain, or into the bushes, or down the manhole in the main street—my people are shy—may have been only your reflection in the glass. My country specializes in such illusions.

2.

Let me propose myself as typical. I walk upright on two legs, and have in addition two arms, with ten appendages, that is to say, five at the end of each. On the top of my head, but not on the front, there is an odd growth, like a species of seaweed. Some think this is a kind of fur, others consider it modified feathers, evolved perhaps from scales like those of lizards. It serves no functional purpose and is probably decorative.

My eyes are situated in my head, which also possesses two small holes for the entrance and exit of air, the invisible fluid we swim in, and one larger hole, equipped with bony protuberances called teeth, by means of which I destroy and assimilate certain parts of my surroundings and change them into my self. This is called eating. The things I eat include roots, berries, nuts, fruits, leaves, and the muscle tissues of various animals and fish. Sometimes I eat their brains and glands as well. I do not as a rule eat insects, grubs, eyeballs, or the snouts of pigs, though these are eaten with relish in other countries.

3.

Some of my people have a pointed but boneless external appendage, in the front, below the navel or midpoint. Others do not. Debate about whether the possession of such a thing is an advantage or disadvantage is still going on. If this item is lacking, and in its place there is a pocket or inner cavern in which fresh members of our community are grown, it is considered impolite to mention it openly to strangers. I tell you this because it is the breach of etiquette most commonly made by tourists.

In some of our more private gatherings, the absence of cavern or prong is politely overlooked, like club feet or blindness. But sometimes a prong and a cavern will collaborate in a dance, or illusion, using mirrors and water, which is always absorbing for the performers but frequently grotesque for the observers. I notice that you have similar customs.

Whole conventions and a great deal of time have recently been devoted to discussions of this state of affairs. The prong people tell the cavern people that the latter are not people at all and are in reality more akin to dogs or potatoes, and the cavern people abuse the prong people for their obsession with images of poking, thrusting, probing, and stabbing. Any long object with a hole at the end, out of which various projectiles can be shot, delights them.

I myself—I am a cavern person—find it a relief not to have to worry about climbing over barbed-wire fences or getting caught in zippers.

But that is enough about our bodily form.

4.

As for the country itself, let me begin with the sunsets, which are long and red, resonant, splendid and melancholy, symphonic, you might almost say; as opposed to the short boring sunsets of other countries, no more interesting than a light-switch. We pride ourselves on our sunsets. "Come and see the sunset," we say to one another. This causes everyone to rush outdoors or over to the window.

Our country is large in extent, small in population, which accounts for our fear of empty spaces and also our need for them. Much of it is covered in water, which accounts for our interest in reflections, sudden vanishings, the dissolution of one thing into another. Much of it, however, is rock, which accounts for our belief in Fate.

In summer we lie about in the blazing sun, almost naked, covering our skins with fat and attempting to turn red. But when the sun is low in the sky and faint, even at noon, the water we are so fond of changes to something hard and white and cold and covers up the ground. Then we cocoon ourselves, become lethargic, and spend much of our time hiding in crevices. Our mouths shrink and we say little.

Before this happens, the leaves on many of our trees turn blood-red or lurid yellow, much brighter and more exotic than the interminable green of jungles. We find this change beautiful. "Come and see the leaves," we say, and jump into our moving vehicles and drive up and down past the forest of sanguinary trees, pressing our eyes to the glass.

We are a nation of metamorphs.

Anything red compels us.

5.

Sometimes we lie still and do not move. If air is still going in and out of our breathing holes, this is called sleep. If not, it is called death. When a person has achieved death, a kind of picnic is held, with music, flowers, and food. The person so honoured, if in one piece, and not, for instance, in shreds or falling apart, as they do if exploded or a long time drowned, is dressed in becoming clothes and lowered into a hole in the ground, or else burnt up.

These customs are among the most difficult to explain to strangers. Some of our visitors, especially the young ones, have never heard of death and are bewildered. They think that death is simply one more of our illusions, our mirror tricks; they cannot understand why, with so much food and music, the people are so sad.

But you will understand. You, too, must have death among you. I can see it in your eyes.

6.

I can see it in your eyes. If it weren't for this I would have stopped trying long ago to communicate with you in this halfway language that is so difficult for both of us, that exhausts the throat and fills the mouth with sand; if it weren't for this I would have gone away, gone back. It's this knowledge of death, which we share, where we overlap. Death is our common ground. Together, on it, we can walk forward.

By now you must have guessed: I come from another planet. But I will never say to you, "Take me to your leaders." Even I—unused to your ways though I am—would never make that mistake. We ourselves have such beings among us, made of cogs, pieces of paper, small disks of shiny metal, scraps of coloured cloth. I do not need to encounter more of them.

Instead I will say, "Take me to your trees. Take me to your breakfasts, your sunsets, your bad dreams, your shoes, your nouns. Take me to your fingers; take me to your deaths."

These are worth it. These are what I have come for.

Time Capsule Found on the Dead Planet

1. In the first age, we created gods. We carved them out of wood; there was still such a thing as wood, then. We forged them from shining metals and painted them on temple walls. They were gods of many kinds, and goddesses as well. Sometimes they were cruel and drank our blood, but also they gave us rain and sunshine, favourable winds, good harvests, fertile animals, many children. A million birds flew over us then, a million fish swam in our seas.

Our gods had horns on their heads, or moons, or sealy fins, or the beaks of eagles. We called them All-Knowing, we called them Shining One. We knew we were not orphans. We smelled the earth and rolled in it; its juices ran down our chins.

2. In the second age we created money. This money was also made of shining metals. It had two faces: on one side was a severed head, that of a king or some other noteworthy person; on the other face was something else, something that would give us comfort: a bird, a fish, a fur-bearing animal. This was all that remained of our former gods. The money was small in size, and each of us would carry some of it with him every day, as close to the skin as possible. We could not eat this money, wear it, or burn it for warmth; but as if by magic it could be changed into such things. The money was mysterious, and we were in awe of it. If you had enough of it, it was said, you would be able to fly.

3. In the third age, money became a god. It was all-powerful, and out of control. It began to talk. It began to create on its own. It created feasts and famines, songs of joy, lamentations. It created greed and hunger, which were its two faces. Towers of glass rose at its name, were destroyed and rose again. It began to eat things. It ate whole forests, croplands, and the lives of children. It ate armies, ships, and cities. No one could stop it. To have it was a sign of grace.

4. In the fourth age we created deserts. Our deserts were of several kinds, but they had one thing in common: nothing grew there. Some were made of cement, some were made of various poisons, some of baked earth. We made these deserts from the desire for more money and from despair at the lack of it. Wars, plagues, and famines visited us, but we did not stop in our industrious creation of deserts. At last all wells were poisoned, all rivers ran with filth, all seas were dead; there was no land left to grow food.

Some of our wise men turned to the contemplation of deserts. A stone in the sand in the setting sun could be very beautiful, they said. Deserts were tidy, because there were no weeds in them, nothing that crawled. Stay in the desert long enough and you could apprehend the absolute. The number zero was holy.

5. You who have come here from some distant world, to this dry lakeshore and this cairn, and to this cylinder of brass, in which on the last day of all our recorded days I place our final words:

Pray for us, who once, too, thought we could fly.

"The Peach Women of Aa'A" from *The Blind Assassin*

In the evenings there's dancing, smooth glittery dancing on a slippery floor. Induced hilarity: she can't avoid it. Everywhere around, the flash-bulbs pop: you can never tell where they're aiming, or when a picture will appear in the paper, of you, with your head thrown back, all your teeth showing.

In the mornings her feet are sore.

In the afternoons she takes refuge in memory, lying in a deck chair, behind her sunglasses. She refuses the swimming pool, the quoits, the badminton, the endless, pointless games. Pastimes are for passing the time and she has her own pastime.

The dogs go round and round the deck on the ends of their leashes. Behind them are the top-grade dog-walkers. She pretends to be reading.

Some people write letters, in the library. For her there's no point. Even if she sent a letter, he moves around so much he might never get it. But someone else might.

On calm days the waves do what they are hired to do. They lull. The sea air, people say—oh, it's so good for you. Just take a deep breath. Just relax. Just let go.

. . . . .

Why do you tell me these sad stories? she says, months ago. They're lying wrapped in her coat, fur side up, his request. Cold air blows through the cracked window, streetcars clang past. Just a minute, she says, there's a button pressing into my back.

That's the kind of stories I know. Sad ones. Anyway, taken to its logical conclusion, every story is sad, because at the end everyone dies. Birth, copulation, and death. No exceptions, except maybe for the copulation part of it. Some guys don't even get that far, poor sods.

But there can be happy parts in between, she said. In between the birth and the death—can't there? Though I guess if you believe in Heaven that could be a happy story of sorts—dying, I mean. With flights of angels singing you to your rest and so forth.

Yeah. Pie in the sky when you die. No thanks.

Still, there could be happy parts, she says. Or more of them than you ever put in. You don't put in many.

You mean, the part where we get married and settle down in a little bungalow and have two kids? That part?

You're being vicious.

Okay, he says. You want a happy story. I can see you won't leave it alone until you get one. So here goes.

. . . . .

It was the ninety-ninth year of what was to become known as the Hundred Years' War, or the Xenorian Wars. The Planet Xenor, located in another dimension of space, was populated by a super-intelligent but super-cruel race of beings known as the Lizard Men, which wasn't what they called themselves. In appearance they were seven feet tall, scaly, and grey. Their eyes had vertical slits, like the eyes of cats or snakes. So tough was their hide that ordinarily they didn't have to wear clothing, except for short pants made of carchineal, a flexible red metal unknown on Earth. These protected their vital parts, which were also scaly, and enormous I might add, but at the same time vulnerable.

Well, thank heaven something was, she says, laughing.

I thought you'd like that. Anyway, their plan was to capture a large number of Earth women and breed a super-race, half-human, half-Xenorian Lizard Man, which would be better equipped for life on the various other habitable planets of the Universe than they were—able to adjust to strange atmospheres, eat a variety of foods, resist unknown diseases, and so on—but which would also have the strength and the extraterrestrial intelligence of the Xenorians. This super-race would spread out through space and conquer it, eating the inhabitants of the different planets en route, because the Lizard Men needed room for expansion and a new source of protein.

The space fleet of the Lizard Men of Xenor had launched its first attack on Earth in the year 1967, scoring devastating hits on major cities in which millions had perished. Amid widespread panic, the Lizard Men had made parts of Eurasia and South America their slave colonies, appropriating the younger women for their hellish breeding experiments and burying the corpses of the men in enormous pits, after eating the parts of them they preferred. They liked the brains and the hearts especially, and the kidneys, grilled lightly.

But the Xenorian supply lines had been cut by rocket fire from hidden Earth installations, thus depriving the Lizard Men of the vital ingredients for their zorch-ray death guns, and Earth had rallied and struck back—not only with her own fighting forces but with clouds of gas made from the poison of the rare Iridis *hortz* frog once used by the Nacrods of Ulinth to tip their arrows, and to which, it had been discovered by Earth scientists, the Xenorians were particularly susceptible. Thus the odds had been evened out.

Also their carchineal shorts were flammable, if you could hit them dead on with a missile that was hot enough already. Earth snipers with bull's-eye aim, using long-range phosphorus-bullet guns, were the heroes of the day, although retaliations against them were severe, and involved electrical tortures previously unknown and excruciatingly painful. The Lizard Men did not like having their private parts burst into flame, which was understandable.

Now, by the year 2066, the alien Lizard Men had been beaten back into yet another dimension of space, where Earth fighter pilots in their small, quick two-man harry-craft were pursuing them. Their ultimate goal was to wipe out the Xenorians entirely, keeping perhaps a few dozen for display in specially fortified zoos, with windows of unbreakable glass. The Xenorians, however, were not giving up without a fight to the death. They still had a viable fleet, and a few tricks left up their sleeves.

They had sleeves? I thought they were naked on top.

Judas Priest, don't be so picky. You know what I mean.

Will and Boyd were two old buddies—two scarred and battle-seasoned harry-craft veterans of three years' standing. This was a long time in the harry-craft service, where losses ran high. Their courage was said by their commanders to exceed their judgement, though so far they had got away with their rash behaviour, raid after daring raid.

But as our story opens, a Xenorian zorch-craft had closed in on them, and now they were shot to hell and limping badly. The zorch-rays had put a hole in their fuel tank, knocked out their link with Earth control, and melted their steering gear, giving Boyd a nasty scalp wound in the process, whereas Will was bleeding into his spacesuit from an unknown site in midsection.

Looks like we're for it, said Boyd. Screwed, blued, and tattooed. This thing's gonna go kablooey any minute now. I just wish we'd of had the time to blast a few hundred more of the scaly sons of guns to kingdom come, is all.

Yeah, ditto. Well, mud in your eye, old pal, said Will. It looks like you've got some running down in there anyway—red mud. Ha, ha.

Ha, ha, said Boyd, grimacing in pain. Some joke. You always had a bum sense of humour.

Before Will could reply, the ship spun out of control and went into a dizzying spiral. They'd been seized by a gravity field, but of which planet? They had no idea where they were. Their artificial-gravity system was kaput, and so the two men blacked out.

When they awoke, they couldn't believe their eyes. They were no longer in the harry-craft, nor in their tight-fitting metallic spacesuits. Instead they were wearing loose green robes of some shining material and reclining on soft golden sofas in a bower of leafy vines. Their wounds were healed, and Will's third finger on the left hand, blown off in a previous raid, had grown back. They felt suffused with health and well-being.

Suffused, she murmurs. My, my.

Yeah, us guys like a fancy word now and then, he says, talking out of the side of his mouth like a movie gangster. It gives the joint a bit of class.

So I imagine.

To proceed. I don't get it, said Boyd. You think we're dead?

If we're dead I'll settle for dead, said Will. This is all right, all righty.

I'll say.

Just then Will gave a low whistle. Coming toward them were two of the peachiest dames they had ever seen. Both had hair the colour of a split-willow basket. They were wearing long garments of a purplish-blue hue, which fell in tiny pleats and rustled as they moved. It reminded Will of nothing more than the little paper skirts they put around the fruit in snooty Grade-A grocery stores. Their arms and feet were bare; each had a strange headdress of fine red netting. Their skin was a succulent golden pink. They walked with an undulating motion, as if they'd been dipped in syrup.

Our greetings to you, men of Earth, said the first.

Yes, greetings, said the second. We have long expected you. We have tracked your advent on our interplanetary tele-camera.

Where are we? said Will.

You are on the Planet of Aa'A, said the first. The word sounded like a sigh of repletion, with a small gasp in the middle of it of the kind babies make when they turn over in their sleep. It also sounded like the last breath of the dying.

How did we get here? said Will. Boyd was speechless. He was running his eyes over the lush ripe curves on display before him. I'd like to sink my teeth into a piece of that, he was thinking.

You fell from the sky, in your craft, said the first woman. Unfortunately it has been destroyed. You will have to stay here with us.

That won't be hard to take, said Will.

You will be well cared for. You have earned your reward. For in protecting your world against the Xenorians, you are also protecting ours.

Modesty must draw a veil over what happened next.

Must it?

I'll demonstrate in a minute. It merely needs to be added that Boyd and Will were the only men on Planet Aa'A, so of course these women were virgins. But they could read minds, and each could tell in advance what Will and Boyd might desire. So very soon the most outrageous fantasies of the two friends had been realized.

After that there was a delicious meal of nectar, which, the men were told, would stave off age and death; then there was a stroll in the lovely gardens, which were filled with unimaginable flowers; then the two were taken to a large room full of pipes, from which they could select any pipe they wanted.

Pipes? The kind you smoke?

To go with the slippers, which were issued to them next.

I guess I walked into that one.

You sure did, he said, grinning.

It got better. One of the girls was a sexpot, the other was more serious-minded and could discuss art, literature, and philosophy, not to mention theology. The girls seemed to know which was required of them at any given moment, and would switch around according to the moods and inclinations of Boyd and Will.

And so the time passed in harmony. As the perfect days went by, the men learned more about the Planet of Aa'A. First, no meat was eaten on it, and there were no carnivorous animals, though there were lots of butterflies and singing birds. Need I add that the god worshipped on Aa'A took the form of a huge pumpkin?

Second, there was no birth as such. These women grew on trees, on a stem running into the tops of their heads, and were picked when ripe by their predecessors. Third, there was no death as such. When the time came, each of the Peach Women—to call them by the names by which Boyd and Will soon referred to them—would simply disorganize her molecules, which would then be reassembled via the trees into a new, fresh woman. So the very latest woman was, in substance as well as in form, identical with the very first.

How did they know when the time had come? To disorganize their molecules?

First, by the soft wrinkles their velvety skin would develop when overripe. Second, by the flies.

The flies?

The fruit flies that would hover in clouds around their headdresses of red netting.

This is your idea of a happy story?

Wait. There's more.

. . . . .

After some time this existence, wonderful though it was, began to pall on Boyd and Will. For one thing, the women kept checking up on them to make sure they were happy. This can get tedious for a fellow. Also, there was nothing these babes wouldn't do. They were completely shameless, or without shame, whichever. On cue they would display the most whorish behaviour. *Slut* was hardly the word for them. Or they could become shy and prudish, cringing, modest; they would even weep and scream—that, too, was on order.

At first Will and Boyd found this exciting, but after a while it began to irritate.

When you hit the women, no blood came out, only juice. When you hit them harder, they dissolved into sweet mushy pulp, which pretty soon became another Peach Woman. They didn't appear to experience pain, as such, and Will and Boyd began to wonder whether they experienced pleasure either. Had all the ecstasy been a put-on show?

When questioned about this, the gals were smiling and evasive. You could never get to the bottom of them.

You know what I'd like right about now? said Will one fine day.

The same thing I'd like, I bet, said Boyd.

A great big grilled steak, rare, dripping with blood. A big stack of French fries. And a nice cold beer.

Ditto. And then a rip-roaring dogfight with those scaly sons of guns from Xenor.

You got the idea.

They decided to go exploring. Despite having been told that Aa'A was the same in every direction, and that they would only find more trees and more bowers and more birds and butterflies and more luscious women, they set out toward the west. After a long time and no adventures whatsoever, they came up against an invisible wall. It was slippery, like glass, but soft and yielding when you pushed on it. Then it would spring back into shape. It was higher than they could possibly reach or climb. It was like a huge crystal bubble.

I think we're trapped inside a big transparent tit, said Boyd.

They sat down at the foot of the wall, overcome by a profound despair.

This joint is peace and plenty, said Will. It's a soft bed at night and sweet dreams, it's tulips on the sunny breakfast table, it's the little woman making coffee. It's all the loving you ever dreamed of, in every shape and form. It's everything men think they want when they're out there, fighting in another dimension of space. It's what other men have given their lives for. Am I right?

You said a mouthful, said Boyd.

But it's too good to be true, said Will. It must be a trap. It may even be some devilish mind-device of the Xenorians, to keep us from being in the war. It's Paradise, but we can't get out of it. And anything you can't get out of is Hell.

But this isn't Hell. It's happiness, said one of the Peach Women who was materializing from the branch of a nearby tree. There's nowhere to go from here. Relax. Enjoy yourselves. You'll get used to it.

And that's the end of the story.

That's it? she says. You're going to keep those two men cooped up in there forever?

I did what you wanted. You wanted happiness. But I can keep them in or let them out, depending how you want it.

Let them out, then.

Outside is death. Remember?

Oh. I see. She turns on her side, pulls the fur coat over her, slides her arm around him. You're wrong about the Peach Women though. They aren't the way you think.

Wrong how?

You're just wrong.

Appendices

An Open Letter from Margaret Atwood to the Judson Independent School District

First, I would like to thank those who have dedicated themselves so energetically to the banning of my novel *The Handmaid's Tale*. It's encouraging to know that the written word is still taken so seriously.

That thought aside, I would like to congratulate the students, parents, and teachers who have supported the use of my book in advanced placement courses. They have aligned themselves against the censurers, book-banners, and book-burners throughout the ages, and have stood up for open discussion and a free expression of opinion—which, last time I looked, was still the American way, though that way is under pressure.

I would also like to comment on the objections to the book that have been made. The remark "offensive to Christians" amazes me—why are some Christians so quick to see themselves in this mirror? Nowhere in the book is the regime identified as Christian. It puts into literal practise some passages from the Bible, but these passages are not from the New Testament. In fact, the regime is busily exterminating nuns, Baptists, Quakers, and so forth, in the same way that the Bolsheviks exterminated the Mensheviks. The only person who says anything Christian is the heroine herself. You will find her own version of The Lord's Prayer at the end of Chapter 30.

As for sexual explicitness, *The Handmaid's Tale* is a good deal less interested in sex than is much of the Bible. Leaving aside the Song of Solomon, there's quite a bit of sex—rape, incest of various kinds, seduction, lust, prostitution, public intercourse on a rooftop with one's father's concubines, and more. One of the things that makes the Bible such a necessary book is its refusal to throw a lace tablecloth over this kind of behaviour.

The sexual point in my book would seem to be that all totalitarianisms try to control sex and reproduction one way or another. Many have forbidden inter-racial and inter-class unions. Some have tried to limit childbirth, others have tried to enforce it. It was a common practise for slave owners to rape their slaves, for the simple purpose of making more slaves. And so on.

The other point would be that the free choice of a loved one—when denied by a regime or a culture—is going to happen anyway, though under such conditions it will be both brave and dangerous. I give you *Romeo and Juliet*. Also, when marriage itself has been made into a travesty, talk of sex within the bonds of marriage becomes simply fatuous.

Two last thoughts. First, I put nothing into my book that human beings have not already done. It's not a pretty picture, but it's our picture, or part of it. Second, if you see a person heading toward a huge hole in the ground, is it not a friendly act to warn him?

Again, I congratulate you, and wish you well. Your thoughtfulness and courage have set an example well worth following.

Sincerely,
Margaret Atwood

Weird Tales Covers of the 1930s

". . . you could have a pack of nude women who've been dead for three thousand years, with lithe, curvaceous figures, ruby-red lips, azure hair in a foam of tumbled curls, and eyes like snake-filled pits . . . I could throw in some sacrificial virgins as well, with metal breastplates and silver ankle chains and diaphanous vestments. And a pack of ravening wolves, extra . . . Popular on the covers—they'll writhe all over a fellow, they have to be beaten off with rifle butts."

These words appear in my 2000 novel, *The Blind Assassin.* They're spoken by Alex Thomas, who's a writer of pulp-magazine fiction in the 1930s. He's not writing at this moment in the novel, however: he's picking up a girl in a park. His initial method is storytelling, always a good thing to know something about, whichever role you're playing. If you're the pickup artist, it's as well to be able to tell a good story or two, and if you're the target you need to be able to determine if you've heard them before.

The fictional Alex Thomas got his beautiful vamps and their adornments straight off the covers of *Weird Tales,* definitely the sort of magazine he'd have wanted to publish in. In the 1930s and '40s, *Weird Tales* published, well, weird tales: fantasy, horror, and sci-fi of the bug-eyed monster variety. Its covers were in lurid colour, lovingly drawn in pastels by Margaret Brundage—the only female pulp cover artist of her era—who was fresh from a career as a fashion designer and illustrator.

Brundage specialized in vicious or threatened young women, sometimes totally nude, but otherwise dressed in colourful and revealing outfits involving metal brassieres, translucent veils, and ankle chains both decorative and functional, often accessorized with whips and shackles. Large fanged animals are a recurring motif: the Brundage women have equivocal relationships not only with wolves but with other charismatic carnivores. Sometimes the women appear frightened by their dangerous friends, but they may also stride forth, alpha females leading the pack.

The Brundage covers run from 1933 through the early 1940s, making them a perfect source for my invention, Alex Thomas; so it's clear where Alex got his clichés. But—looking back at these clichés now—I wonder where I myself got them? I wasn't born when Brundage was creating most of her covers; yet her subject matter seems very familiar to me. When you're a child, you soak up images like a sponge. It doesn't matter to you where they come from. In those timeless years between infancy and, say, seven, what is has always been: in that way, children inhabit the realm of myth.

In the 1940s, when I was a comic generation kid, there were certain things we all knew. We took it as a given that children could make friends with wolf packs, and might even be raised by them; these packs would rush to their aid in times of peril. I had my own imaginary pack of this kind, and therefore was not alarmed by Al Capp's Wolf Gal of the popular 1940s cartoon strip *L'il Abner.* Wolf Gal must have been the first Brundage-like carnivorous pinup I ever saw. She had white hair and fierce white eyebrows, she most likely ate men, she was scantily dressed, and like all the members of Capp's harem of eccentric glamour gals (stunners such as Stupefyin' Jones, Appassionata Von Climax, and the mud-covered pig-fancier Moonbeam McSwine), she was what was once called "bountifully endowed." *Hubba hubba*, men said in those days: a term obscure in origin but most likely a variant of *hübsche*, the German word for "beautiful."

Books and characters in books, pictures and elements in pictures—they all have families and ancestors, just like people. What generated Wolf Gal? Probably Brundage's wolf gals of *Weird Tales*, which—I'd bet—Capp would have read and drawn from. Was their grandparent Kipling's *The Jungle Book*, in which the wolf-raised child was a boy? Did these clawed lovelies devolve from the high art of the late nineteenth century, so fond of depicting femmes fatales paired with animals to show how animalistic they were underneath? Or does the line stretch way back, to folklore and tales of lycanthropy, or even further back, to times when animals were thought to assume human form at will?

The enduring popularity of werewolf stories must be based on something, and that something may be close to a wish. Was Margaret Brundage, unknown to herself, drawing early versions of that trope of female freedom, women who run with the wolves? Bram Stoker, the author of *Dracula*, was neither the first nor the last to supply seductive women with canine teeth somewhat larger than is generally desirable in a girlfriend. (It's to be noted that Wolf Gal has no Mr. Wolf Gal, and we strongly suspect that Wolf Gal—like some furry Turandot or a female spider—has been the death of all lovelorn aspirants to her hand, or paw.)

Then there are the women in the twin tinnies—those two

shiny cups, attached to the torso with fine chain link—that abound in Brundage's oeuvre. Richard Wolinsky produced a recorded documentary called *The Girl in the Brass Brassiere: An Oral History of Science Fiction 1920–1950*, a title that acknowledges the ubiquity of the trope in early twentieth-century sci-fi and fantasy, but like everything else pictorial, this item of clothing had its visual predecessors.

The message borne by the hard-but-soft frontage is mixed. One part of it derives from orientalism. Before moving to *Weird Tales*, Margaret Brundage drew covers for another pulp, *Oriental Stories*. In the exotic maidens she portrayed, Brundage was lifting from a rich vein of nineteenth-century Victorian orientalist painting, some of it purporting to depict such things as harems and girl-slave markets, but some of it purely imaginative, inspired by the hugely influential *One Thousand and One Nights*. This iteration of the metal bra—non-functional, skimpy, and bejewelled—invokes bondage and/or other depravities. Robert E. Howard of *Conan the Barbarian* fame—a frequent publisher in *Weird Tales*—was quite keen on both slave girls and depravities, and used the Brundage dress code. In *The Blind Assassin*, I based Alex Thomas's writhing women with eyes like snake-filled pits on simple-hearted Conan's encounters with the uncanny seductresses of the corrupt, decaying cities through which he marauds.

Brassiere advertisements from the 1940s and 1950s hint at the second part of the twin-tinnie lineage: impermeability. Maidenform was just one of the brands featuring blindingly white bras with concentric circles of stitching that suggested armour. Their ads that coupled a state of undress with public activities—"I dreamed I was a private eye in my Maidenform bra"; "I dreamed

I was a lady editor in my Maidenform bra"—presented the bra less as an aid to seduction than as a guarantee of security and, combined with the name, of chastity. Athene, the maiden goddess, with her shield and spear and her helmet, is perhaps a distant relative.

A closer relative is the Valkyrie, a virgin demi-goddess from Norse mythology whose job was to gather up dead warrior-heroes and cart them off to Odin's banquet hall. Richard Wagner brought the Valkyries to the opera stage in his *Ring Cycle*, but to a 1940s and 1950s audience they were more familiar as the parody conception of what a Wagnerian soprano should look like: large metal brassiere or corset, long braids, helmet complete with Viking-fantasy wings. Sure enough, there's Bugs Bunny in the 1957 cartoon *What's Opera, Doc?*, cross-dressing as the Valkyrie Brünnhilde, with pink-winged helmet and two tiny brass cups stuck on his chest.

Wonder Woman, the comic-book heroine who first appeared in 1941, doesn't have the full metal jacket, but she does have enough shiny stuff on her front to indicate her lineage. She, too, is related to the virgin goddesses—the chaste moon-goddess Artemis, in her case. Supergirls of all kinds, good and bad, are generally unmarried: Wonder Soccer Mom, amazing though she may be in real life, somehow doesn't quite fit the image.

The metal bra was capable of carrying two simultaneous undermeanings: vulnerability, especially when it was flimsily attached to a girl with big, scared eyes; or strength and staunch resistance, when the "breast plates," as they were called in the pulps, were more substantial and their wearer looked determined. Brundage sometimes tried for both at once: a girl in a brass brassiere and little else, with big, scared eyes, tiptoeing forward with fear but determination, anklets quivering, to unlock some handsome fellow from a cage.

The "low art" of one age often cribs from the "high art" of the preceding one; and "high art" just as frequently borrows from the most vulgar elements of its own times. The *Lady Chatterley* porno-trial wars were fought over whether several words you could see scribbled on a washroom wall every day had the right to be written inside something that purported to be "literature." The *Weird Tales* covers of the 1930s are just one example of the way cultural memes transmit themselves, taking their meaning in part from their context and from our own knowledge of it. Thus, from Wagner's ultra-serious Valkyries to Brundage's equivocal brass bras, to Maidenform's faux-naïve undergarments, to Bugs Bunny's skimpy travesties, and finally to Madonna's witty pop-show quotation of the entire tradition. And from the wolf-women of myth and folklore to Brundage's wolf-girls, to Al Capp's gloss on them in his *L'il Abner* Wolf Gal, to me as child reader, and finally to my invention, Alex Thomas.

Alex is using *Weird Tales* pulp schlock as foreplay. He knows it's schlock, and the girl he's seducing knows it as well, but that's part of the attraction, for her as well as for him. "I don't think I could fob those off on you," he says of the depraved women and the maidens in sexual peril he's conjuring up for her. "Lurid isn't your style."

"You never know," the girl replies. "I might like them."

And so she does.

Acknowledgements

Many thanks to the following, who made the Ellmann Lectures so enjoyable for me:

Joseph Skibell, director of the Richard Ellmann Lectures in Modern Literature; Barbara Freer Skibell; Sharon Hart-Green, associate professor of Near and Middle Eastern Civilizations, University of Toronto; Michael P. Kramer, professor of English, Bar-Ilan University; and Esther Schor, professor of English, Princeton University. Members of the Emory administration: James W. Wagner, president of the university; Earl Lewis, provost and executive vice president for academic affairs; Rosemary M. Magee, vice president and secretary of the university; Robin Forman, dean of Emory College of Arts and Sciences. Alicia Franck and Tom Jenkins, Becky Herring, Nicholas Surbey, Levin Arnsperger, and the many members of the Emory University faculty and staff who also contributed.

I would also like to thank my agents, Phoebe Larmore and Vivienne Schuster; my editors, Ellen Seligman of McClelland & Stewart, Canada; Nan Talese of Nan A. Talese/Doubleday, U.S.A.; and Lennie Goodings of Virago Press, U.K. Also: my copy editor, Heather Sangster; John Shoesmith and Jennifer Toews of the Thomas Fisher Rare Book Library; the Judith Merril Collection at the Toronto Public Library; and the Widener Library at Harvard University. Thanks to all publishers who have granted permission for some of the earlier pieces, and to the many newspaper and magazine editors with whom I have worked over the years. Finally, thanks to my office staff, Sarah Webster, Anne Joldersma, Laura Stenberg, and Penny Kavanaugh. And to the many writers included in this book, whose work I have enjoyed over the course of sixty-odd years.

Permissions Acknowledgements

Grateful acknowledgement is made to the following for permission to reprint previously published material:

The three chapters in Part 1 are based on the Richard Ellmann Lectures in Modern Literature, delivered at Emory University, Atlanta, Georgia, 24–26 October 2010.

Woman on the Edge of Time by Marge Piercy. Originally published as "An Unfashionable Sensibility" in *The Nation*, 4 December 1976, pp. 601–2. *Second Words: Selected Critical Prose 1960–1982* by Margaret Atwood (Toronto: House of Anansi Press, 1982), pp. 272–78. Reprinted with permission of the publishers.

H. Rider Haggard's *She* (New York: Random House, 2002), pp. xvii–xxiv. *Moving Targets: Writing with Intent, 1982–2004* by Margaret Atwood (Toronto: House of Anansi Press, 2004), pp. 234–41. *Curious Pursuits* by Margaret Atwood (London: Virago Press, 2005), pp. 249–56. *Writing with Intent: Essays, Reviews, Personal Prose 1983–2005* by Margaret Atwood (New York: Carroll & Graf Publishers/Perseus Books, 2005), pp. 198–204. Reprinted with permission of the publishers.

The Queen of Quinkdom: *The Birthday of the World and Other Stories* by Ursula K. Le Guin. *The New York Review of Books*, Vol. 49, No. 14, 26 September 2002. *Curious Pursuits* by Margaret Atwood (London: Virago Press, 2005), pp. 297–308. *Moving Targets: Writing with Intent, 1982–2004* by Margaret Atwood (Toronto: House of Anansi Press, 2004), pp. 281–92. *Writing with Intent: Essays, Reviews, Personal Prose 1983–2005* by Margaret Atwood (New York: Carroll & Graf Publishers/Perseus Books, 2005), pp. 243–53. Reprinted with permission of the publishers.

Arguing Against Ice Cream: *Enough: Staying Human in an Engineered Age* by Bill McKibben. *The New York Review of Books*, 12 June 2003. *Moving Targets: Writing with Intent, 1982–2004* by Margaret Atwood (Toronto: House of Anansi Press,

2004), pp. 339–50. *Writing with Intent: Essays, Reviews, Personal Prose 1983–2005* by Margaret Atwood (New York: Carroll & Graf Publishers/Perseus Books, 2005), pp. 294–304. Reprinted with permission of the publishers.

George Orwell: Some Personal Connections. An address broadcast on the BBC Radio 3 on 13 June 2003. Reprinted as "Orwell and Me," *Guardian*, 16 June 2003. *Moving Targets: Writing with Intent, 1982–2004* by Margaret Atwood (Toronto: House of Anansi Press, 2004), pp. 331–38. *Curious Pursuits* by Margaret Atwood (London: Virago Press, 2005), pp. 333–40. *Writing with Intent: Essays, Reviews, Personal Prose 1983–2005* by Margaret Atwood (New York: Carroll & Graf Publishers/Perseus Books, 2005), pp. 287–93. Reprinted with permission of the publishers.

Ten Ways of Looking at *The Island of Doctor Moreau*, by H. G. Wells (London: Penguin, 2005). *Curious Pursuits* by Margaret Atwood (London: Virago Press, 2005), pp. 383–96. *Writing with Intent: Essays, Reviews, Personal Prose 1983–2005* by Margaret Atwood (New York: Carroll & Graf Publishers/Perseus Books, 2005), pp. 386–98. Reprinted with permission of the publishers.

Never Let Me Go by Kazuo Ishiguro. Originally published as "Brave New World: Kazuo Ishiguro's novel really *is* chilling" in *Slate* magazine (www.slate.com), 1 April 2005. Reprinted with permission of *Slate* magazine.

After the Last Battle: *Visa for Avalon* by Bryher. *The New York Review of Books*, Volume LII, No. 6, 7 April 2005.

Brave New World by Aldous Huxley (Toronto: Vintage Canada, 2007). Reprinted as "Everyone Is Happy Now," *Guardian*, 17 November 2009.

Of the Madness of Mad Scientists: Jonathan Swift's Grand Academy. From *Seeing Further: The Story of Science & The Royal Society*, ed. Bill Bryson (London: HarperPress, 2010), pp. 37–48.

Cryogenics: A Symposium. Originally published in *When the Wild Comes Leaping Up: Personal Encounters with Nature*, ed. David Suzuki (Vancouver: Greystone Books, 2002), pp. 143–47.

Cold-Blooded. From *Good Bones* by Margaret Atwood (Toronto: Coach House Press, 1992), pp. 65–70. (London: Virago Press, 1993), pp. 65–70. (Toronto: New Canadian Library/McClelland & Stewart, 1997), pp. 53–56. Also published in *Good Bones and Simple Murders* by Margaret Atwood (Toronto: McClelland & Stewart, 1994), pp. 79–83. (New York: Nan A. Talese/Doubleday, 1994), pp. 79–83. Also published in *Bones and Murder* by Margaret Atwood (London: Virago Press, 1995), pp. 85–90. Reprinted with permission of the publishers.

Homelanding. From *Good Bones* by Margaret Atwood (Toronto: Coach House Press, 1992), pp. 121–28. (London: Virago Press, 1993), pp. 121–28. (Toronto: New Canadian Library/McClelland & Stewart, 1997), pp. 91–96. Also published in *Good Bones and Simple Murders* by Margaret Atwood (Toronto: McClelland & Stewart, 1994), pp. 132–38. (New York: Nan A. Talese/ Doubleday, 1994), pp. 132–38. Also published in *Bones and Murder* by Margaret Atwood (London: Virago Press, 1995), pp. 141–47. Reprinted with the permission of the publishers.

Time Capsule Found on the Dead Planet. Printed in *Guardian*, 26 September 2009.

"The Peach Women of Aa'A." From *The Blind Assassin* by Margaret Atwood (Toronto: McClelland & Stewart, 2000), pp. 349–56. (London: Bloomsbury Publishing, 2000), pp. 349–56. (New York: Doubleday, 2000), pp. 349–56. Reprinted with the permission of the publishers.

An Open Letter from Margaret Atwood to the Judson Independent School District. Originally printed in the *San Antonio Express-News*, 12 April 2006.

Weird Tales Covers of the 1930s. Originally published in *Playboy*, September 2011.

A NOTE ABOUT THE AUTHOR

Margaret Atwood's books have been published in more than thirty-five countries. She is the author of more than forty books of fiction, poetry, and critical essays. In addition to *The Handmaid's Tale,* her novels include *Cat's Eye,* shortlisted for the Booker Prize; *Alias Grace,* which won the Giller Prize in Canada and the Premio Mondello in Italy; *The Blind Assassin,* winner of the 2000 Booker Prize; *Oryx* and *Crake,* shortlisted for the 2003 Booker Prize; and her most recent, *The Year of the Flood.* She lives in Toronto with writer Graeme Gibson.